EARN IT

Life, Times, and Leadership
by
Sergeant Major of the Army
Julius William (Bill) Gates

and
Colonel Wes Martin

Introduction by Governor Thomas J. Ridge, former United States Secretary of Homeland Security

THE NCO HISTORICAL SOCIETY
TEXAS, USA

Title: EARN IT
Subtitle: Life, Times, and Leadership by Sergeant Major of the Army Julius William (Bill) Gates
Authors: Julius W. Gates and Wes Martin
Published by: The NCO Historical Society

© 2025 by Julius W. Gates and Wes Martin

All rights reserved. No part of this book may be reproduced or transmitted in any form or by any means, electronic or mechanical, including photocopying, recording or by any information storage and retrieval system, without written permission from the author, except for the inclusion of brief quotations in a review.

Contact the NCO Historical Society for changes or updates at info@ncohistory.com.

All photographs are the property of the author, unless otherwise noted.

Cover photograph courtesy United States Army
Sergeant Major of the Army rank insignia image courtesy The Institute of Heraldry.

Library of Congress Control Number: 2025931809

ISBN-13: 978-0-9963181-7-4

Published in the United States of America

1st Edition: July 2025

CREED OF THE NONCOMMISSIONED OFFICER

No one is more professional than I. I am a Noncommissioned Officer, a leader of Soldiers. As a Noncommissioned Officer, I realize that I am a member of a time-honored corps, which is known as "The Backbone of the Army." I am proud of the Corps of Noncommissioned Officers and will at all times conduct myself as to bring credit upon the Corps, the Military Service and my country regardless of the situation in which I find myself. I will not use my grade or position to attain pleasure, profit, or personal safety.

Competence is my watchword. My two basic responsibilities will always be uppermost in my mind – accomplishment of my mission and the welfare of my Soldiers. I will strive to remain technically and tactically proficient. I am aware of my role as a Noncommissioned Officer. I will fulfill my responsibilities inherent in that role. All Soldiers are entitled to outstanding leadership; I will provide that leadership. I know my Soldiers and I will always place their needs above my own. I will communicate consistently with my Soldiers and never leave them uninformed. I will be fair and impartial when recommending both rewards and punishment.

Creed of the Noncommissioned Officer

Officers of my unit will have maximum time to accomplish their duties; they will not have to accomplish mine. I will earn their respect and confidence as well as that of my Soldiers. I will be loyal to those with whom I serve; seniors, peers, and subordinates alike. I will exercise initiative by taking appropriate action in the absence of others. I will not compromise my integrity, nor my moral courage. I will not forget, nor will I allow my comrades to forget that we are professionals, Noncommissioned Officers, leaders!

RANGER CREED

Recognizing that I volunteered as a Ranger, fully knowing the hazards of my chosen profession, I will always endeavor to uphold the prestige, honor, and high esprit de corps of the Rangers.

Acknowledging the fact that a Ranger is a more elite Soldier who arrives at the cutting edge of battle by land, sea, or air, I accept the fact that as a Ranger my country expects me to move further, faster and fight harder than any other Soldier.

Never shall I fail my comrades. I will always keep myself mentally alert, physically strong and morally straight and I will shoulder more than my share of the task whatever it may be, one-hundred-percent and then some.

Gallantly will I show the world that I am a specially selected and well-trained Soldier. My courtesy to superior officers, neatness of dress and care of equipment shall set the example for others to follow.

Energetically will I meet the enemies of my country. I shall defeat them on the field of battle for I am better trained and will fight with all my might. Surrender is not a Ranger word. I will never leave a fallen comrade to fall into the hands of the enemy and under no circumstances will I ever embarrass my country.

Readily will I display the intestinal fortitude required to fight on to the Ranger objective and complete the mission, though I be the lone survivor.

Rangers lead the way!

ACKNOWLEDGEMENTS

"Earn It" were the last two words spoken by Tom Hanks' character Captain John Miller in the now-classic World War II motion picture *Saving Private Ryan*. This comment caught retired Sergeant Major of the Army Julius William Gates' attention as it expressed the philosophy he had practiced throughout his entire life. Bill, as he was known to his friends, served as the Army's top soldier ten years prior to our meeting. A professional bond between us developed as soon as we met, not just over soldiering, but over a mutual respect of history. As we were both living in Albuquerque at the time, we visited every national park, historical site, and anything related to military history in New Mexico.

Sometimes the drives took hours and between us on the console was a recorder taping our conversations. Those tapes captured discussions about leadership, training, soldier development, team building, and combat. Bill's knowledge was unfathomable. In one conversation, Bill casually referred to the long-term result of military operations in Borneo. It became obvious that despite all of the information we had captured on tape, there was a vast reservoir of knowledge that needed to be further explored and documented. The overwhelming majority of his knowledge came not from the classroom, but from experience and independent study.

Bill was receptive when I suggested that we put his experience and knowledge into a book where it could be made available to future generations. Naming the book *Earn It* was his idea. Our subsequent discussions expanded from weekends on the road to evenings at his kitchen table. The conversation part was easy; then came the arduous work of bringing it all together. Again, at the table we reviewed all the information; determining what should be dropped, remain, expanded, and included in the narrative. I quickly realized that Bill is not a prima donna and dislikes anything focused solely on him. The family values, work ethic, and love of his nation that developed during his early years on the North Carolina farmlands remain with him to this day.

Sergeant Major Gates' success was not just a result of his upbringing and personal values. Once in the United States Army he was blessed by the influence of great leaders. Our goal in this book was not only to recognize them, but more importantly to explain how their leadership styles had a long-term impact on their subordinates who went on to share those lessons with the next generation.

History is not stagnant, every day it progresses forward. Gates witnessed that progression and was a part of it. The US Marines have an expression that should be used by all branches of the armed forces: "improvise, adapt, and overcome." Bill's ability to exercise those three words allowed him to develop with history as it too was developing. We conscientiously chose not to just write about Sergeant Major Gates' life, the places he served, and ongoing events that brought him to those places. That would only have covered the who, what, where, and when. Left out would have been the why and how. Those are critical elements of any type of analysis.

Fortunately, the "Be All You Can Be" motto has been brought back into service. Those five words do not just pertain to individual soldiers; they apply to every team and every command. *Earn It* is our contribution to help achieve that objective.

Appreciation is extended to Judy Bowyer Martin for the hundreds of hours provided in transcribing dozens of tape-recorded interviews and multiple proofreadings as each chapter was developed. Codie Martin's computer technical assistance work throughout the project and creating the index was invaluable. The proofreading and advice provided by Marnie Salter and Colonel (US Air Force, Retired) Lowell Little were critical in developing the final text. Appreciation is also extended to Lieutenant General (US Army, Retired) Jack Gardner, Colonel (US Army, Retired) Bill Ivey, and former Pennsylvania Governor Tom Ridge for their reviews, guidance, and recommendations. Special recognition is provided to Steven and Leya Booth of Genius Book Services for editing, cover, and interior design, and to CSM (Ret.) Daniel Elder and our publisher, the NCO Historical Society.

Wes Martin
Colonel, United States Army (Retired)

FOREWORD

Earn It removes all doubt that leaders are made, not born. Julius William (Bill) Gates earned his way through life to become the Eighth Sergeant Major of the Army at a critical time in our Nation's history. America has always been able to have great leaders rise to answer the call to duty. This is not just with generals like George Washington and Colin Powell, it's also with sergeants like Alvin C. York and Bill Gates.

The making of leaders is not merely through the individual's effort. The work ethic instilled in Sergeant Major Gates while growing up on a farm is evident. Coming from a family and community where all members looked out for each other prepared him well for military service. In the Army, Sergeant Major Gates was once again in an environment where people stood firmly beside each other and united their efforts to achieve common goals for their commands and for the nation they served. His determination to study and train hard as a young soldier was recognized by his sergeants and officers. Not only did his seniors encourage him to attend advanced training, but they also helped him prepare as they did with all soldiers under their supervision.

At every step of life's journey Sergeant Major Gates accepted challenging positions. This further contributed to his progression in experience, knowledge, and rank. As he rose in seniority, he shared his knowledge with subordinates, peers, and seniors. His skills and professionalism put him in hot demand. In every command he served as top soldier, the next higher commander recognized Bill Gates as a valuable asset. His rapid progression serving as battalion, division-forward, and division sergeant major within three years is remarkable.

From the Berlin Brigade to racial integration of America's universities, Vietnam, the Cold War, Operation JUST CAUSE in Panama, and Operation DESERT SHIELD/STORM in the Middle East, *Earn It* tracks many of America's experiences through much of the second half of the 20th century. Sergeant Major Gates provides a boots-on-the-ground perspective. His insights will never be found in an academic narrative. The "Spirit of the Bayonet" chapter is perhaps the best analysis of a warrior's mindset that has ever been written.

Having myself served as an infantry staff sergeant in Vietnam I saw the effects that Asian War had both on our service members and our country. Panama and Iraq in 1991 and early 1992 did not repeat mistakes of the previous generation. Anchored on the Weinberger-Powell Doctrine, our country's leaders ensured all the right decisions were made and resources were in place before offensive ground operations commenced. Clear attainable objectives were identified up front, risk and consequence analyses were conducted, exit strategies were determined, diplomatic options were exhausted, and support of both American and international populations was achieved.

Two topics covered by Sergeant Major Gates are very special to me. The first is teamwork from squad level to interservice and coalition cooperation. No one branch of the US armed forces is more important than the other.

Earn It emphasizes the criticality of working together and respecting each other.

The second topic is the importance of family in the military. Sergeant Major Gates was blessed with a wonderful wife who not only looked out for him but looked out for others. Margaret ensured young spouses received proper support. When she visited an orphanage of Korean children fathered by American service members, Margaret was outraged at the squalid conditions. Like all children, they had no control over the nature of their birth. It was Margaret's actions that brought an end to the neglect.

Sergeant Major Gates was just a young soldier when General Douglas McArthur gave his acceptance speech upon receiving the Sylvanus Thayer Award at West Point in 1962. In that speech, General McArthur stated, "Duty, Honor, Country: Those three hallowed words reverently dictate what you ought to be, what you can be, what you will be. They are your rallying points: to build courage when courage seems to fail; to regain faith when there seems to be little cause for faith; to create hope when hope becomes forlorn." *Earn It* is a soldier's testimonial of those words.

Sergeant Major Gates has taken time to share with us many of the lessons he learned, providing the historical perspective and importance of what was happening while he was serving his nation. Furthermore, he not only explains how individuals can improve themselves, but also how teams and commands can do the same.

Earn It is more than an autobiography. It is a soldiers' guide and a leadership manual for all branches of the military. It's also an excellent example of what can be accomplished when dedicated members of the armed forces team together. For many years I have witnessed the joint effort of Sergeant Major Bill Gates and Colonel Wes Martin. This book should be on the recommended reading list of every general officer and command sergeant major.

Thomas J. Ridge
43d Governor of Pennsylvania
First US Secretary of Homeland Security
Former Staff Sergeant, 23d Infantry Division, US Army

Contents

CREED OF THE NONCOMMISSIONED OFFICER iii
RANGER CREED v
ACKNOWLEDGEMENTS vii
FOREWORD x
CHAPTER 1   Growing up in North Carolina 1
CHAPTER 2   Making of a Soldier 13
CHAPTER 3   Mentoring from Professional Leaders 30
CHAPTER 4   Impact of Great Leaders 47
CHAPTER 5   Baptism by Fire 62
CHAPTER 6   Assessment and Reflections of the Vietnam War 77
CHAPTER 7   Return to the States 94
CHAPTER 8   Spirit of the Bayonet 107
CHAPTER 9   The Harder Right 121
Photos: 138
CHAPTER 10   Be All You Can Be 148
CHAPTER 11   Ranger Company First Sergeant 157
CHAPTER 12   Garlstedt 175
CHAPTER 13   Third Infantry Division
               Command Sergeant Major 197
CHAPTER 14   Sergeants Major Academy 215
CHAPTER 15   US Forces Korea Command Sergeant Major 228
CHAPTER 16   Sergeant Major of the Army 244
CHAPTER 17   Low and High Intensity Conflicts 263
CHAPTER 18   Epilogue 287
APPENDIX   Professional Soldiering 289
Acronyms 308
About the Authors 310
Index 311

CHAPTER 1
Growing up in North Carolina

In America, the nature of one's birth is not an ultimate factor in life's future achievements. General Omar N. Bradley was born the son of a farmer and rural schoolteacher in Moberly, Missouri. General Colin L. Powell, born in Harlem, was raised in the Bronx. His parents had emigrated from Jamaica and worked blue-collar jobs to support their family. The ultimate factor in their life's achievements was that both had parents who passed on the importance of hard work and academic study. Forty years apart, both men would rise to serve as Chairman of the Joint Chiefs of Staff during critical times in American history.

Despite humble and different beginnings, a person can succeed in life if a persistent willingness exists to "Earn it." This is the basis of the Army slogan that, at the time of this writing, has been reactivated. "Be All You Can Be" applies to individual soldiers, squads, platoons, and command levels all the way to the top. Emphasizing the necessary relationship between commanders and the troops they lead, the man who led the relief of Bastogne and later served as Army Chief of Staff, General Creighton W. Abrams, Jr., stated, "A general may plan battles, but he cannot advance very far without soldiers." Mine is a soldier's story that originated in the mountains of North Carolina and spanned five decades in uniform. It worked its way through Germany, Vietnam, numerous American bases, the Pentagon, and continues to this day. I am no longer wearing the uniform, and my dedication to our nation and the United States military is no less than when I first enlisted. Born six months before the 7 December 1941, Japanese attack on Pearl Harbor, I was too young to understand the full impact of World War II while it was going on. That changed as I matured, and time progressed.

History tells us the Great Depression ended with America's entry into the war. This is partially true. In the North Carolina hills it took longer for the impact of those difficult years to conclude. We were raised on a farm about five miles northwest of Carrboro, North Carolina, next to Chapel Hill, where the state university is located. My mother, Ethel Lee Douglas, was born to a full-blooded Cherokee woman. My father, Jack Alvin Gates, came from an Irish-English combination. We were a self-made family, looking for no handouts, believing in hard work, and earning whatever was received.

Going without was a fact of life that our parents taught us to accept. During my earliest years, our family lived in a very small house on a farm with a living room that doubled as a bedroom for my mother and father, a small kitchen, and an upstairs that was an open bedroom for all the kids. Dignity and family integrity prohibited us from trying to beg, bum, or steal items for us to live in the society in which we were part. None of us ever crossed the line. Parental discipline would have made it a one-time event.

Running water was not an in-house luxury. Cooking, laundry, and bathing required carrying water from a small spring about three to four hundred meters to the house. Bathing in colder weather was inside the house in a transportable tin tub. During warmer weather, we bathed in the creek. Winter or summer did not matter when we relieved ourselves in an outhouse. A barn housed the cows and a mule. On the right side of the smokehouse was a small storage garage. Looking back, by today's standards, it could be considered pretty rough. While growing up in that environment, we kids did not think it was that bad. It taught us unity and living together with a common purpose. We also knew one important fact. Our parents were working hard to make our lives better. With all of us pitching in together, they succeeded.

For me, it was perfect preparation for an army career starting in the 1950s. The transition from civilian to military life was very easy. Open bay barracks and an austere way of life came as no shock. More important than the physical aspects of being raised on the family farm was the mental maturity that came with it, especially the recognition that hard work and application defined our character. My family was a tight-knit group of five boys and four girls. I do not want to use the word "clan," but it was close to that. Being older, sister Alice, brother Thomas, and my parents were my early role models. Their impact contributed immensely to the building of my work ethic, determination to succeed against the odds, commitment to take care of those around me, and to look out for those who were placed in my care. Whatever income Alice made; she shared with all of us. When I completed elementary school, Alice bought me a jacket and a tie for the graduation ceremony. When somebody tried to hurt me, Thomas would step in.

That farm was our livelihood. North Carolina's climate and soil allowed us excellent crops of beans, peas, potatoes, tomatoes, and just about every other type of vegetable. Fresh vegetables were greatly enjoyed during harvest season, but we always planned ahead for the winter and months that followed before the next harvest. Canning foods during late summer and into autumn was a family event.

In addition to crops, we also raised chickens and hogs, both for ourselves and for selling. There is only so much variety that comes from chicken. The hogs were a different story, providing a selection of pork, ham, bacon, and sausage. Long before vegetable oils became the health standard, lard from pigs and hogs was the universal cooking grease.

For us to live, we had to *earn it*. Other people in the community did receive federal government assistance, like free cheese and other food items. Not only did my parents never accept anything our family did not earn, but they flat out refused to have anything to do with it. Game hunting provided us with further variety. Of the shotgun and the .22-caliber rifle, I preferred the rifle. Knowing the entire family would be able to enjoy what was brought home increased my determination to make every shot count. Learning and then applying the basics of sight alignment, sight picture, grip, stance, breathing, proper trigger squeeze, and not anticipating the weapon's discharge allowed me to take down a squirrel or rabbit at one hundred yards. The shotgun required less aiming skill, but getting so much closer to the target was more likely to alert the animal. Many good mentors were involved in my initial weapons training, including military veterans, my uncles, and especially my brother Thomas.

Our only automobile was a 1935 Ford pickup truck. To my mind, this was the most enduring and versatile vehicle ever produced. It hauled everything: hay, spare parts, and the family. Henry Ford built his vehicles to be fuel-efficient and durable. The reliability of that vehicle was also enhanced by the excellent care and maintenance it received from my parents. Lessons learned on their care of that truck would also play an important role for me in later life. Many times growing up, we would go to school without shoes to wear. Some kids who were a little better off always made fun of us. It was not just the Gates family, but a whole bunch of families. We knew what we were doing and knew of our parents' continual efforts to make things better for us. We were patient.

In my family, during World War II, one brother served in the Philippines, and two uncles elsewhere in the Pacific. Two more uncles served together in Europe. My mother's brother, Sonny Douglas, was a combat medic who participated in the Normandy invasion. During the assault, he worked non-stop to save lives. Months later, after slugging through France and entering Germany, he was a witness to the most despicable and horrifying act of inhumanity imaginable. Sonny was part of the spearhead that liberated Nazi extermination camps. A look at the photographs he brought back was evidence of the extent of the atrocities committed against Jewish people. Once back home, Sonny did his part in ensuring General Dwight D. Eisenhower's goal of educating as many people as possible about what the Hitler regime had done. Eisenhower correctly assessed that one day there would be people who would deny that the Holocaust occurred.

Through my schooling, I learned that Americans fought in the Revolution to bring independence to their future nation. Through those photographs and with Sonny's words, I learned that in World War II, Americans fought for humanity and to end the brutal tyranny cast upon the world. Even though I was still young then, the experiences shared with us from the returning combat veterans were very influential in creating my desire to one day be a soldier.

By the time Uncle Harvey shipped out for Korea, I was old enough to understand war. Harvey was more like a brother because he had lived in that small house with us and was part of our direct family. From uncles who had deployed to World War II combat, Harvey received much guidance. The influence from the rest of us encouraged Harvey to do his best and return home alive. Harvey succeeded on all counts.

School teachers played a critical role during the wars. Many of us, like my family, living in the country, did not have electricity. Listening to the radio or watching the developing technology of television were not options. Our teachers filled the gap. They provided the historical perspective and kept us informed as the war progressed.

Our teachers were also our champions. While in fifth grade, I was riding a bicycle when I was struck by an automobile. This resulted in a fractured skull, two weeks in a coma, and missing three months of school. Determined not to remain in the fifth grade for a second year, I completed all the requirements to advance with my peers with the help of my sister Alice and my teacher. The school principal objected, saying, "There is no way this boy could miss three months and complete all his requirements." Miss Jones won the fight, informing the principal, "This boy has earned the right to advance to the sixth grade."

Cut from the same cloth as Miss Jones was my third-grade teacher, Miss Agnes Andrews. She was adamant that all her students learn mathematics. Because of her, I can still say the multiplication table, and that's amazing. She was strong on learning the basics, then building from there. If you talked in class, she had a ruler. She would either pop you on the hand or pop you on the head to stop you from talking. We never took it personally, but we did stay quiet.

If Mr. James were not the finest high school teacher I had, he would have been close to it. He not only taught us from the history book, but also from his experiences in Korea. More than a teacher, he was someone to emulate. He carried himself with dignity, respected peers and students alike, maintained discipline in his classroom, and ensured that we all learned. One of the statements he made when we first started his 9th grade class was, "You already have an A in this class. Now you must work to keep it." That made sense to me. In my mind, I said, "I am good enough to have an A and I'm going to keep it." The result was straight

A's in history, with Mr. James' philosophy carrying over to geography, science, agriculture, and physical education. All straight A's.

My worst subject was algebra; I just could not get interested in it, mainly due to a lack of understanding of what I would do with it later in life. Every day I have lived since high school has been one without the need to use algebra. Looking back, my teachers, Miss Jones, Miss Andrews, and Mr. James, were outstanding leaders. All ensured we learned the basics, built upon the basics, and maintained firm discipline in the ranks. None of them hesitated a second to go into the long fight in our defense. The lessons I learned from these three teachers stayed with me my whole life in both academic and professional leadership.

My earlier-mentioned bicycle accident had another critical impact on me. Few expected me to live, let alone fully recover. Being temporarily out for the count, I did not have any say about it until I came out of the coma. In the long run, aside from some hearing loss in my left ear, it was a full recovery. Maybe something else rattled in my head. After returning from the expected dead, reasonable fear of heights had disappeared. Years later, I liked climbing on cliffs only to rappel back down. From bad events often come good results if the resolve exists to push through.

There was a tender impact of that accident as well. When I woke up, the first person I saw was my mother. According to my sisters, our mother spent ninety percent of her time in that hospital waiting for me to wake up and come back home. A mother's love and tenderness are gifts that last a lifetime.

When the Korean War finished, like World War II, our veterans returned proudly wearing uniforms. Part of this was pride, boosted by the warm receptions they received throughout their journeys home. The other part was that they often did not have the chance to buy civilian clothes until their return. Once they were home, we held special events to honor them. Not just our families, but other people within the county turned out. We did not have parades, but lots of individual and sometimes group parties. Whether in community halls or outdoors in good weather, there was never a shortage of good cheer and good food that community mothers prepared.

In the early 1950s, my father put his auto maintenance skills to use further. He purchased a service station and a repair shop along State Highway 54. It was about three miles from our farm, one mile by dirt road and two more by the highway's pavement. Operating a farm and a service station meant more responsibility and work for our family. We kids had matured in age and behavior to the point that our parents could rely on us to share the workload with ever-decreasing supervision. My family built a large house about a quarter mile from the service station. Constructed from the cellar up, the first floor consisted of a large living room, a two-part kitchen, and a bedroom for our parents. On the second floor were two large bedrooms, one for the boys and one for the girls. Anxious to move in, we accepted, still carrying water to the house and taking baths in that tin tub until complete indoor plumbing was set up. The service station also broadened my awareness. Compared to the gravel routes they replaced, those two-lane blacktop and cement roads, like Highway 54, were the superhighways of their time. For my family, providing customer service allowed us to visit with people from all directions. The interstate system, which Eisenhower established while president based upon his observations of the German autobahns, was still decades away from reaching all of America.

Including with the service station was a grocery store, small by today's standards. The community's character was reflected in how people paid for their groceries. Even though most charged for some groceries and gasoline, they were determined to settle their accounts. My father did not have to run after them or send them letters. On Friday, they received their paychecks. On Saturday morning, they were at our station paying bills. Occasionally, they would have a problem, a family emergency, possibly losing their job, and couldn't pay. Nobody ever went around looking for them to collect. Many times, my father threw the charge into the trash and said, "Just let it go." My father had not forgotten the difficult times of his own family's life. That was country living where people trust, work together, and look out for each other.

During harvest season, the community rotates around each other's farms in the evenings. There was always lots of fun, good conversations, and plenty of food. For us kids, the best was the boiled corn covered with melted butter. The elderly menfolk usually had the best time, often off somewhere, consuming their corn from a moonshine jar. In the country, what some people would consider a boring occurrence, we turned into a celebration. The community also came together when a member had just suffered from a drastic situation, or needed help. All of the other families would come in and assist if someone's hay could not be brought in from the field before rain and could suffer from rotting. There was never any money exchanged. We all knew the supported family would assist another family in the future. Support and friendship were never selective; race and religion were never factors. We worked, celebrated, and shared each other's pain during difficult times.

Even though we had to work hard, we had our share of fun. One night, my brother Thomas and I crawled out of the woods to where a religious revival was being held inside a pavilion covered by a huge tin roof. Just when the preacher was yelling about the immediate coming of the Lord, Thomas threw a rock onto the roof. The initial hit and the bouncing of the rock on the tin created a series of bangs. Whether he believed it or was ad-libbing in the moment, the preacher screamed, "Oh, Lord, I told all you. That is the coming of the Lord—on top of this building tonight—so you better pay attention and listen." We stayed hidden and never told anyone about our involvement that special night when the Lord ascended to the top of that tin-roof pavilion, deep in the North Carolina hills.

Another night, returning from the theater, Thomas and I were walking down the road and came across piles of manure. When cow patties bake in the sun, they dry from the outside, in. The outside can be hard, but the inside is just as soft as when it dropped out of the cow. We took a bunch of those hard-shell patties and laid them two rows deep across the highway. The color of the patties and the darkness of the road blended into the nighttime very well. We only had to hide in the bush briefly before we saw a car coming. The patties exploded when crushed by the tires. The driver drove maybe seventy-five feet and stopped the car. Thinking he had at least one flat, with a flashlight in hand, the driver commenced checking his tires by rubbing his hand over the rubber. Probably not realizing he had been pranked, but with a palm full of wet cow manure, the driver commenced into a tirade of cussing. Thomas and I stayed quiet and did not leave our hiding place until the driver departed.

When they became of age, my older brothers, Thomas, Wallace, and Roland, went out independently. My father left my younger brother, Jimmy, and me to work with the family business. The service station was rented out to Dave Crabtree, allowing us to place primary emphasis on the farm. I was just the right age to pick up the passion. Being the oldest son at home and having a job at the Ford dealership, I now had the responsibility to run the cross-breeding hog farm under my father's supervision. Becoming part of Future Farmers of America allowed me to enter our hogs in different county contests throughout the region.

Being a certified hog farmer, my father was authorized to pick up food garbage at the Chapel Hill hospital and the university and bring it back. We had a huge vat where we would cook the garbage for a certain amount of time and temperature per regulations. Eating that garbage prepared those hogs for quick sale. Two hundred and fifty pounds is a top hog. We made good money from the hog business.

Each success provided us with the financial resources to further build the farm. One of the most significant upgrades was the purchase of a tractor, which provided relief for the man behind and the mule in front of the plow. Land not dedicated to livestock was used to grow wheat, oats, barley, and corn.

The only time I did not tell my father the entire truth concerned that tractor and the plowing of a field during my teen years. Coming up towards Saturday, a friend of mine, Edgar Lord, and I wanted to go to the movies that night with some girls we had met. Normally, by six or seven o'clock, I would be finished feeding all the cows and all the other work. This Saturday, I also had three acres of land to plow. I realized there was no way in the world I would finish in time, do all that stuff, and be able to go to the movie. So, on Friday, I told Edgar my predicament. He said, "I'll tell you what . . . my dad and mom are gone for the weekend. I'll come over there tomorrow with my dad's tractor and plow. With both plows we'll get done awful quick."

In a couple of hours, that land was plowed. I drove the tractor back up to the new house we lived in. When I pulled in, my father asked me, "What's wrong? Is something wrong with the tractor?" I informed him the ground was plowed. He said, "What're you talking about? Ain't no way in hell you can plow that much ground that quick." I repeated that I was finished and needed time off to go to a movie. We got in the truck and rode down.

He looked at it and said, "Damn, I've never seen anything like that. How in the world did you plow that much ground in that length of time?" As it turned out, he gave me five dollars, and we had a ball that night. Years later, when he talked to people, he would point to me and say, "See that boy right there? That's the plowingest damn boy in the whole county . . . the whole Orange County . . . right there! He can plow more land in a short period of time than anyone else around here can." I never told him the whole story.

An irritant to all farmers in the state was the federal government restriction forbidding us to grow more than two acres of cotton. This was when the Department of Agriculture was determining where and in what quantity certain crops would be grown. Limiting the amount of land farmers could use to grow cotton ultimately drove us completely out of the market. Cotton must be marketed in mass quantities to make a profit. Limiting the amount of acreage meant that raising cotton was not worth the effort. Even if all the farmers in the area pooled their crops, it was still not enough to compete with corporate businesses.

In the end, the upbringing of all the Gates children worked out for the best. During those early days, we were poor in one sense, but not in the other. All my brothers and sisters became successful in life because of the work ethic that was ingrained into us.

CHAPTER 2
Making of a Soldier

As far back as memory can go, I always wanted to join the US Army. The experiences shared by the returning combat veterans of the Korean War, the unbiased explanations provided by our schoolteachers, and the conversations within our homes contributed to that desire. Throughout America, the critical purpose of the United States military was understood and appreciated. Even service members passing through to distant destinations were treated with complete respect. Being seventeen years of age at the time, my enlistment on August 12, 1958, required my parents' signature. Everyone gets excited when traveling into the unknown, and I was no different. With a bus ticket provided by the recruiter, I left Chapel Hill and went to an in-processing station in Raleigh. The most critical and life-impacting part of in-processing was the Oath of Allegiance, to swear to protect the Constitution of the United States against all enemies (foreign and domestic) and obey those in authority over me. There was no way of knowing, with certainty, exactly where that oath would take me. What I did know was my own determination to live up to the words just sworn.

The next bus trip was with everyone who had raised their hands. We were going to Fort Jackson, South Carolina and another in-processing center. There, we were issued our equipment, including uniforms, boots, sanitation items, bedding, and field gear. Our training sergeants made it clear that everything we needed was in that training issue and to accept our new names, which they immediately provided—"trainee" and "boot." Even though our pay grade was E–1, a "private," we were not allowed to use that title. That honor and being called "soldiers" were withheld until we proved ourselves by graduating from basic combat training (BCT). There was an interesting learning curve. Those from the cities and suburbs had to learn about getting up before 0600 (6 a.m.), shooting firearms, hard physical work, and taking orders. My hardest part was understanding why we were sleeping so late. As for the physical part, a backpack was a lot lighter than a bale of hay, a bucket of hog slop, and a sack of wheat. In the country, we have a saying, "Don't get above your raisin'." My upbringing was living with the land. Having my request to be in the infantry approved allowed me to build upon my existing skills.

We did not have professionally schooled sergeants to guide us through our training, just individuals who were detailed for that round of BCT. In addition to the standard utility uniform, they wore a helmet on their head and a pistol belt around their waist. Some of them were very professional, but not all. We never saw the first sergeant. The commander appeared only for Saturday morning inspections.

This lack of oversight created a perfect storm for abuse of authority. One trainee by the name of Haseck had a rough time. Two of the sergeants took it upon themselves to physically beat him hard. That stayed with me for my entire career. There were other ways to deal with this soldier. Physical contact solves nothing. In time, I would learn just how much these beatings were in violation of the Uniform Code of Military Justice. Those sergeants got away with the beatings because we did not know any better, and senior company leadership was absent. The longer I remained in the ranks, the more experience I gained, the more it led me to become increasingly hostile to abuse of authority, especially at the expense of a subordinate or anyone who could not defend themselves.

Except for those abusive sergeants, we did have outstanding noncommissioned officers (NCOs) in BCT and later in Advanced Individual Training (AIT). They did the best they could, especially considering they were detailed in the training mission while not professionally prepared. Future implementation of the drill sergeant program was designed to fill this gap. The result is that today's professionally trained drill sergeants are much better prepared to understand their responsibilities and to develop new recruits into soldiers. Unfortunately, there will always be unprofessional and self-serving people in all positions of authority. An absent chain-of-command will encourage abuse of authority, as was cast upon Trainee Haseck.

Two of my greatest pleasures in basic training was having my own clothes and being issued a .30–caliber M1 rifle. The clothing issue was simple. Growing up in a large family meant that all the clothes you had were temporary. You got them when an older sibling outgrew them and passed them on to a younger sibling when you grew too big.

Until then, the M1 was the finest weapon I had ever fired. The reputation it earned in World War II was legendary. The German infantry carrying bolt-action rifles was no match for Americans with rapid fire capabilities. The M1's accuracy at 200 meters was phenomenal. Adapting my rifle and shotgun skills gained in the North Carolina hills to such a fine and durable weapon was no problem. The rifle was so tough that it was nearly unbreakable.

Many of the trainees had grown up in an urban or suburban environment. They had to work harder but did come through, proving that the nature of one's birth and rearing are ultimately not the determining factors in how much they can achieve in life. The ones who wanted to succeed in my basic training platoon worked hard, applied themselves to being team members, and listened closely to their instructors.

Eight weeks after the time we began our training, we marched by the pass-and-review stand at our graduation ceremony. We had passed our first mission and had earned the right to be called "private" and "soldier." Following a two-week break, I returned to Fort Jackson for advanced individual training (AIT) to be transformed into an infantry soldier. The training was built upon the land navigation skills I received in BCT, and I learned how to establish a route by using a map and compass. Our firearms training advanced to larger .30– and .50–caliber machine guns. Then came indirect fire, which involved firing mortar shells from tubes and calling in artillery by providing map coordinates to a point on the ground where we wanted the artillery guns to land their rounds. Calling for fire gave us the appreciation for extreme accuracy and made sure we got it right and ensured that we did not call artillery on top of ourselves or someone else. With AIT completed in late December 1958, we were transported by train to Fort Dix, New Jersey, which served as our staging area. In early January, we were bused to the shipping port and placed on a troop ship bound for Germany. The cross-Atlantic trip took seven days because we went from New Jersey down to North Carolina and picked up some additional soldiers from the 82d Airborne who were going to Europe for their next assignments.

Even if no one was in uniform for that trip, it would have been easy to tell the US Navy crew from the Army passengers. The Navy guys were the ones able to keep their meals in their stomachs. For me, pulling kitchen police duty (working in the kitchen) the entire trip prevented the slightest thought that this was a luxury ocean cruise. Arrival in the North Sea

German port of Bremerhaven was very welcome, but adjusting from a rolling deck back to solid ground took a few hours.

Come evening, we were placed on the Berlin Duty Train, the US Army's rail connection between West Germany and West Berlin. Rather than just one, it was four trains. Each night, two would depart from Berlin en route to Bremerhaven and Frankfurt in central Germany. At the same time, trains from Bremerhaven and Frankfurt would depart for Berlin. The tracks west of Berlin and east of Helmstedt, West Germany, were through East German territory.

The East German road system for automobiles was a few miles away from the rail line. In passing through East Germany, Western military members were officially hosted by the Soviet Army. When our train stopped at Marienborn, East Germany, Soviet guards came on board to check our paperwork and walk through the train. We had already been briefed not to pay the Soviets much attention but not to cause an incident. Our train passed through East Germany without a problem, and we arrived in West Berlin.

Ground access to Berlin, both military and civilian, was critical for both logistical support of the city and to ensure the transportation corridors remained open. At the end of World War II, Berlin and all of Germany were divided into four sectors, each controlled by the governments of the United States, Britain, France, and the Soviet Union, respectively. The divisions were supposed to be only until the stability of Germany could be achieved. The Soviet Union reneged on its part of the agreement, while the three Western powers allowed their occupied territories to be united back into one city and one country. In June 1948, Soviet dictator Joseph V. Stalin had intended to strangle West Berlin by closing off all ground transportation. President Harry S. Truman raised the stakes on Stalin by creating the Berlin Airlift. For fifteen months, using 250,000 flights, allied planes delivered critical supplies to civilians and military personnel in West Berlin.

Stalin, having personally dealt with Truman at the end of World War II during the Potsdam Conference, and knowing Truman had already approved the dropping of two atomic bombs on an enemy of America, and given the size of coalition forces still maintained in Western Europe under the command of General Eisenhower, Stalin knew better than order the shooting down of allied aircraft. The Cold War was on and remained so even after Stalin died in 1953. Through Soviet politburo power plays, Nikita S. Khrushchev became the Soviet boss of bosses. Meanwhile, Eisenhower had finished serving as Commander, Supreme Headquarters Allied Powers Europe, and had been elected US president. The presence of Eisenhower, the old war horse, in the White House gave Soviet leadership pause when thinking about aggression directed against America or its interests.

US-led coalitions had stalemated Communist aggression against the West. The front lines were the North-South Korean border and the East-West German border. The Korean Demilitarized Zone and Berlin would have been the first to be attacked if hostilities had broken out. A strong military presence and wise diplomacy blocked communist expansion into Western Europe and northwestern Asia. Communism would have to seek a later time and a different part of the world to exploit its aggression.

Against this backdrop of international intrigue, I began my tour of duty inside Berlin. At the time, my mission was to serve as a junior infantryman in A Company, 3rd Battle Group, 6th Infantry. All newly assigned personnel meant staying on base for thirty days. It was a standard restriction and well advised. It gave us time to get acclimated. Berlin was still an open city. Our leadership did not need new assignees wandering around unaware, getting into trouble, or stumbling around East Berlin. During the initial training period, a supervised group tour for newly assigned personnel was provided to East Berlin. We saw Soviet and East German soldiers and civilians going about their daily business. It was a very tense situation and an informative orientation. Even though only fourteen years had elapsed since the division of Germany, it was already becoming obvious the difference in progress between the Soviet sector and that once occupied by France, Britain, and the United States. West Berlin was on the way to becoming a thriving democracy-led metropolis. East Berlin was progressing much more slowly. East Berlin civilians also saw the difference and were quietly resettling in West Berlin to build a better life.

After the thirty-day restriction passed, we were allowed outside the compound and could even go into East Berlin. The latter privilege, I never had a reason to exercise. Outside the compound, we had to stay in our uniforms. This was not just for identification, but to continually remind ourselves to maintain professional behavior. We were not only defenders of, but guests within, our host country. Also, Soviet and East German agents always sought an allied service member to exploit. By professionally maintaining ourselves and attempting to protect ourselves from espionage, we were sending an unspoken message that communist agents needed to look elsewhere for their pigeons. We also sent the message that our professionalism and dedication to our country were not traits their nations would like to go up against on a future battlefield. As soldiers, in every way possible, we were our nation's best deterrent against future conflict. There was no way of knowing that thirty years later I would apply this lesson I learned as a private.

One of my biggest fortunes was our squad leader. He took good care of us. He would listen any time we had a problem. He always prepared us well for inspections by the platoon leader. The key thing was that he really assisted us. He did not come around to grab our clothes and things and throw them on the floor; instead, he would always show us a better way of storing our clothes and arranging our foot lockers so that they met some very high detail standards. Even in cleaning the latrines (communal bathrooms), he would come around and help us, even if it meant scrubbing the floor.

Building on what we had learned in advanced training at Fort Jackson, our squad leader refined our M1 rifle skills in using the Browning Automatic Rifle and applying direct mortar fire. None of these were merely classroom lessons, but hands-on applications. He taught us how to calibrate the mortar, co-aim, and adjust fire. The actual shooting had to wait until going to southern Germany for live fire training.

All American units in Berlin had packed training schedules. The NCOs continually provided training in urban warfare tactics, first aid, communications, weapons skills, and drill and ceremonies. In the First World War, General John J. Pershing coined the phrase, "Give me a soldier who can shoot and salute," our leadership in Berlin made Pershing's phrase a reality. It not only built confidence in ourselves, but it also sent yet another message to our adversaries that coming at us would be very expensive for them. Our company commander and senior NCOs understood the importance of readiness. The senior NCOs were Korean War veterans, and the commander was a Bataan Death March survivor. During the first year of World War II, the Bataan Death March followed the Allied surrender to Japanese forces in the Philippines. Over 70,000 American and Filipino prisoners of war were forced to walk sixty-five miles while subjected to abuse, starvation, and dehydration. Prisoners who collapsed were executed on the spot by their Japanese captors.

The Berlin Brigade was the pilot project for what would become military operations in urban terrain (MOUT), and we were proud of it. The tactics, techniques, and procedures (TTPs) used by today's Army and Marine Corps started in Berlin. In demolition training, we learned how to set explosive charges on buildings, bridges, tunnels, and sewer systems. Timing was just as important as location and the types of charges. Dropping a building on enemy troops or a bridge while it is being crossed is much more effective than just simple destruction.

The mock training village was right next to the communist sector. Every way a building could be taken and defended was practiced right there. Whenever any American, British, or combined combat team went into that village, communist observers were watching us with their binoculars. We were never allowed to talk to Soviet and East German military personnel. They were our target audience, and we wanted them to report to their bosses what we were doing.

Another type of training we did in Berlin was riot control training at the squad, platoon, and even company level. We learned as we progressed in the training. At first, we would line up on an objective and run a frontal assault. That would have worked on a conventional battlefield, but not against a city's population we were supposed to protect in the first place. That would have played right into Soviet hands. We then learned how to maneuver to an objective. The command anticipated that if the Soviets tried to conduct aggression against Berlin, they would first create civil unrest inside the city. There was a precedent. Hitler had done it two decades earlier in the Sudetenland region southeast of Germany before he invaded. The Soviets had also done it to justify overthrowing the Czarist government. Occasionally, the Soviets would do a show of force, usually at Checkpoint Charlie, the crossing point between West and East Berlin. We would get the alert and respond in kind. The Soviets and East

Germans thought they were harassing us. Instead, they gave us a sense of purpose and helped us refine our skills. In Berlin, we had two integrated fire battle groups. The infantry and tank troops would respond together, knowing the artillery was out of sight but very much on the ready. After each standdown, we conducted an after-action review (AAR) to determine what happened, what we did right, what we did wrong, and how we could do it better next time. The Soviets and East Germans had no idea how much they contributed to our professional development.

The Berlin Wall did not start out as the cement structure the West came to know decades later. In 1959, the communists built a fence around West Berlin. There was a neutral zone, and if East German people wandered into it from their side, they risked being shot by communist guards. We were not authorized to shoot at the East German guards to stop them from killing the refugees. If they got across the neutral zone and over the fence, then they were welcomed by West Berliners. Part of the problem East German authorities had was that not all guards were willing to shoot the refugees, despite knowing the severe punishment they would soon endure. The smarter thing for the guards to do was to shoot but intentionally fail to hit their supposed targets. They could later claim bad marksmanship.

Checkpoint Charlie was only for Soviet, French, British, and American government use. There were multiple checkpoints for the Germans, with strict authorizations, to pass through. Workers crossed in both directions each day. Insurance that East Germans would return when business or work was completed generally rested in the fact that their families were still trapped inside East Berlin. But even that hostage situation failed to stop a slow migration of intellectuals and workers into West Berlin.

Quickly after Nazi Germany fell, the Soviets constructed a war memorial in Berlin. When the city was divided between the four allied nations, this memorial was situated across the Soviet line inside West Berlin. As part of the agreement, the Soviets were allowed to maintain and guard the memorial. As part of their show of force, with some of their sharpest and most disciplined soldiers, the Soviets guarded the memorial twenty-four hours a day. We were allowed to go to the site and observe it. Discipline on our part was mandated. It was a memorial to warriors who died in the service of their country. They deserved our respect

Our battle group commander was Colonel Fred C. Weyand, who eventually became Army Chief of Staff. With us, he had already refined his ability to see and feel what was developing around him. Later in Vietnam, he was able to use his skills to recognize that a major attack was coming against the Americans, and so he moved his forces closer to Saigon. That preparation turned the Tet Offensive into a tremendous North Vietnamese defeat. Years later, North Vietnamese Army (NVA) commander General Vo Nguyen Giap acknowledged that during the Tet Offensive, the American military had decimated his forces.

Weyand's successes in Vietnam and beyond were in the future. For now, he was our highly respected Berlin Brigade commander. Every Monday morning, he led battle group physical training (PT). On a huge parade field, we did the "daily dozen" exercises in cadence (together) with the training leader on a wooden stage. We always started with side-straddle hops (jumping jacks) and worked through the other eleven. After the dozen, we broke down into company formations and ran long-distance. One Monday morning, Weyand jumped up on the stage in a true example of leadership. To be heard across that entire parade field, and by every soldier on it, the last thing he needed was a megaphone. He informed us that he had to go to a meeting, and because he could not do PT that morning, neither would anyone else. He dismissed us, and we went back to the barracks. Not just because of the excusal from PT that morning, but everything about Fred Weyand made him one of the finest soldiers to ever serve in the United States military.

Command emphasis was placed on physical fitness and, especially, team sports. If there were a sport, commands had a team for it. Our specialty was football, and we had a not-so-secret weapon. The company executive officer was First Lieutenant Wood. As a West Pointer, he knew how to play and coach football. Wood was a team builder and player, and arrogance was never one of his traits. He put us through all the drills, and we dominated the brigade. Instead of "Lieutenant," on the field we called him "Coach." After Berlin, the next time I saw Coach Wood was in the Pentagon. He was a two-star general, and I was Sergeant Major of the Army. When he saw it was me, he let out a spontaneous surprise yell and hugged me. Sometimes, we did not worry about military decorum.

Our break from Berlin came when we traveled to southern Germany for training at Wildflecken (we called it Violdflecken), Grafenwoehr, and Hohenfels. Convoying there was not a straight shot. We first had to travel west, departing Berlin through Checkpoint Bravo, en route to Checkpoint Alpha at Helmstedt. Years later, the West German-financed autobahn through East Germany allowed the trip to be a smooth ride. In the early 1960s, we spent over half a day bouncing in and out of potholes on two-lane roads. In West Germany, we turned south and found the roads much less painful.

We went through live fire, squad, platoon, and company training at the training sites. This was actual infantry-type tactical training. With blank ammunition, we did some of that type of training in Grunwald Park in Berlin. At those massive southern German training sites, we could use live ammunition in our fire and maneuver exercises while engaging silhouette targets. At those same sites, tankers and artillerymen could put live rounds into their tubes and refine their steel-on-target skills.

The relationship with British soldiers was outstanding. Perhaps it was a language issue, as the French stayed by themselves. As for the British, we trained and conducted joint exercises together. That bonded us and encouraged camaraderie after duty hours. Similarly, the British liked a good beer and to have fun. They tended to laugh more than the Americans. However, they were very serious when it came to training and preparing to fight the next war.

Our senior officers and NCOs had served alongside the British in World War II and had not forgotten how important both nations were to each other. One shortage we never experienced was war stories from either our American or British seniors. The biggest difference in the memories of war was that the British had experienced the bombings and rocket attacks on top of their cities and families. From the home front, they knew the horror of war much more thoroughly than the Americans. That relationship with the British lasted throughout my entire military career. It's the same thing even today. The Americans and the British are very strong allies. No one knows for sure who said it first, but the old adage is correct: "We are two nations divided by a common language." Add the Canadians, Australians, and New Zealanders into that mix, and we have five nations that will stay permanently in allegiance to each other.

One of the most enjoyable things about the Berlin Brigade was that we did not have kitchen police duty. Our focus was always on training and being prepared to engage the enemy immediately. During the first part of my Berlin tour, which was made even more possible because of all the rifle racks in our platoon bays. Once the lock key was brought forward from the platoon sergeant's room and we received ammunition, we could roll out of our beds and be ready to fight. Near the end of my Berlin tour, the weapons were secured in company-level arms rooms.

Attending the Berlin Brigade NCO Academy changed my outlook on the military. In the past, I had been a good soldier and could make a commander's orderly, which was a special position of guard duty soldier of the day. This is how I received many of my overnight passes. Uniform appearance and the questions I was asked to be selected as the orderly were tough. That was partly because we were the Berlin Brigade and so many other soldiers were competing for it. Even when I was not on guard duty, I was studying for it. The reward was a three-day pass that, when issued, was reduced to a one-day pass that expired at midnight. Only ten percent of the platoon could be on pass at any time, including married soldiers. If a three-day pass fully allowed, then other soldiers would have been denied pass privileges for two nights. Just taking one night was all part of load-sharing and team play.

I had already achieved the rank of specialist fourth class (paygrade E–4), but I didn't understand much beyond my role and responsibilities. That's where the NCO Academy came in. This course was not an established Army leadership school. Nonetheless, it proved very valuable for my development. We were taught how to train soldiers, to organize our lesson plans, and to present classes. Our trainers were not "detailed" to the academy, which usually means the units send the soldiers they can do better without. These trainers were selected for their positions because they were among the best of their peers. Using the military phrase "Be, Know, and Do," these guys had it all. Determined to meet their standards, I worked as hard as I could and graduated top of the class. Upon returning to my unit, I was assigned as fire team leader and was ready for the challenge.

Another accomplishment of mine was passing the General Educational Development test. Having joined the Army before finishing high school created a feeling in me of leaving something unaccomplished. One of the best things military bases have to this day are education centers, with people ready to help. It is up to soldiers to make use of them. One afternoon, I went to visit the center. A lady explained the test and let me take it right then. Normally, that test would have been used as a calibration tool to determine what study courses were necessary for successful retesting. I passed it and walked out of the building with my high school diploma equivalent.

The first dislike I developed in the Army was additional collections on payday. In the early 1960s, after taxes and other deductions, monthly pay for junior enlisted ranged between fifty and seventy-five dollars. That came out to be less than two and no more than two and a half dollars a day to spend over the next month. We had to stand in line and report to the pay officer to be paid. Once paid, we would turn to leave and face multiple desks manned by NCOs who were also in the commander's office. Each one would be collecting "donations" for various charities, ranging from Army Emergency Relief to paying for a new company lounge. Each one wanted four, five, or ten dollars. Command pressure did not leave us much of an option because commanders and sergeants were judged by their superiors on how much they could raise in donations. That was just ripping off a young soldier who was already not making much money. In time, the more NCO stripes I added to my uniform, the more outspoken I became about that practice. When I became a first sergeant, there were no arguments about donations. As top soldier, I made sure my soldiers walked out of the commander's office with the money they received from his desk. Any further solicitations for donations would be made at a different time and the soldiers would only be offered the chance to participate. For any of my seniors who did not like it, that was their problem. Apparently, many other soldiers rising in the ranks saw it the same way. Pressure solicitation in the pay office was eventually prohibited by Army-wide regulation.

The second objection I had was the promotion selection process within command levels. The company commander had the promotion authority up to E–4. NCO promotions progressively advanced up the chain of command, with division commanders having E–7 (sergeant first class) and above authority. The negativity of this system outweighed the positive. Primed for abuse, the system allowed promotions based on friendship and involvement within certain organizations. Fellow members of these organizations would promote each other. Too frequently, merit, capabilities, and hard work were bypassed as inferior soldiers advanced. The promotion system needed to be decentralized and reformed. Establishing qualification standards and skills qualification testing greatly improved promotion criteria, but it was still a few years away at the time of my initial assignment in Europe.

Before 1958, the highest enlisted rank was master sergeant (pay grade E–7). That year Congress established the enlisted grades of E–8 and E–9. Initially, the first promotions were to master sergeant (or first sergeant). A year later, select soldiers were promoted to pay grade E–9 and used the longstanding title of sergeant major. It was still hard to move up in rank, as all the more senior positions were occupied. A specialist fourth class could be doing the job of a fire team leader, a sergeant, or a promotion. Still, promotions to that rank were denied because the person filling the slot at that rank might be on special detail somewhere else on base, or at a senior command headquarters causing soldiers to have to perform duties at levels above their rank.

Intending to complete just one tour of duty, I departed Berlin after two years. The trip was in reverse of the way I came—duty train to Bremerhaven and on a ship to New Jersey. In December 1960, I reported to Fort Devens, Massachusetts, where I spent the remaining six months of my enlistment. At Fort Devens, we had a piecemeal infantry brigade.

The battalion I was assigned to was at sixty percent strength. Even though I was only a specialist, I was made the leader of a six-person squad. Much of our time was spent pulling details for the Special Forces intelligence organizations stationed there.

Another detail was honor guard duty. We had a lot of funeral details where we would fire a gun salute for the fallen comrade and fold an American flag that was removed from the top of the coffin of the deceased. At this time in America, the funerals primarily were for veterans from the Spanish-American War era, and the increase of World War I veterans passing was just beginning. The family members and guests treated us very well and with immense respect.

This respect went well beyond funeral details. East Massachusetts was the cradle of American independence. We were welcome wherever we went. Patriotism in those communities was not just seen at the parades and other frequent events. It was felt every day and on every street. If someone ever tried to burn a flag in those communities, the local citizens would certainly have dealt with them.

Finishing my term of service at Fort Devens, I went home and returned to civilian life. There was great satisfaction and confidence in knowing the challenge of serving my nation had been successfully answered. There was greater confidence and satisfaction that went far beyond me. It was all across the country and continued to grow as it had from the beginning of the 20th century.

CHAPTER 3
Mentoring from Professional Leaders

Years earlier, I watched returning Korean War veterans stepping off the train on their return home to North Carolina. They were proud of their uniforms and of their service to our nation. Having spent most of my tour of duty in West Berlin and the remainder in the cradle region of the American Revolution, I had been afforded two of the most ideal assignments possible to respect the importance of the United States military. My mother took custody of my dress and utility uniforms, hanging each in its own see-through plastic bag for safekeeping. She was proud of those uniforms and ensured no harm would come to them.

The first light of the first day back meant getting out and finding employment. Farm work was still there, but that was an evening and weekend responsibility. My first shot at a steady paycheck meant visiting my old boss at the Ford dealership. The good thing about leaving on the best terms possible and for the right reason, to serve our nation, allowed me to walk through the dealership's door into a friendly environment. Within minutes, I was re-hired and promoted to parts department manager.

Good leadership principles easily adapt between military and civilian environments. The person who uses or fails to use those principles is responsible for his or her own success or failure. What I learned on the farm and in the Army, refined in the Berlin Brigade and NCO academy, allowed me to work well with the two employees also assigned to the parts department. They responded with mutual respect, knowing I would already be there when they arrived in the morning and would remain after they finished in the evening.

After being in the Berlin Brigade during the Cold War, stress within the parts department was non-existent. In time, especially with the combat tours yet to come, I learned that ninety percent of an employee's or subordinate's stress comes from the supervisor's behavior. Stress didn't exist in the parts department. It was easy to keep total accountability of the parts, know where they were, order common parts that were continually required, special order unique parts, work with the mechanics to ensure they were able to satisfy customer needs, and assist walk-in customers. Later, in the 1960s, the Army developed the Push System, which was the pre-positioning of equipment necessary to maintain future operations. We had already mastered that concept at that Ford dealership on a much smaller scale.

The dealership was open on Saturday mornings. When it closed, I would stay behind for a couple of hours, making sure everything was in order and preparing for the next week. I was not the only one who came in early and stayed up late. The senior manager was doing the same thing. He would come to me and say, "Bill, it's time to go home." In part, I believe his work ethic prohibited him from having an employee working while he was relaxing. By chasing me out, he could get some rest, knowing I would be getting the same. He would continually encourage me to attend school, especially company training, which would make me a more skilled employee with certifications to prove it. He did not hide his appreciation for the loyalty and dedication I had to both him and his business. That was a two-way street, as I respected his loyalty and trust in me. He kept telling me I should one day have my own business. If I had listened and stayed there, I would probably have owned the dealership. That was his long-term intent.

But there was something missing inside me. It was not that the friends I had grown up with had moved on to jobs in other towns, gotten married, and were attending college. They were friends from my former life. I enjoyed our time together and occasionally seeing them when I returned home. The bonding and camaraderie that develop in the military are different from anything in civilian society. That camaraderie is that you help each other and do not mind doing it. You don't mind putting in extra hours, especially when you're in a demanding environment or having to serve in harm's way.

At the Ford dealership, even though I admired my boss, enjoyed working with everyone, and appreciated being part of a positive environment, I knew what was missing. My manager knew it, fully understood, and was supportive when I told him I needed to return to the ranks. It didn't seem like it then, but I'd made a fast turn-around. Second enlistment came seventy days after being discharged. Once again it was back to Fort Jackson, South Carolina, for in-processing. As I volunteered, the next stop was Fort Campbell, Kentucky, to Company A, 187th Infantry, 3rd Battle Group. The motto of the command was "Rock Solid," and every squad within it had no problem living up to it.

Author Stephen E. Ambrose captured the division's spirit and combat experiences in his book *Band of Brothers*. When I signed into the 101st in the early 1960s, we still had a lot of these senior officers and NCOs from the *Band of Brothers* era. They also served in the Korean War, during which the 187th made the only combat jump. Led then by Colonel William C. Westmoreland, the 187th had the only regimental commander who did a combat jump since World War II. During my first six months in the

101st, now a Major General, Westmoreland, was serving as our division commander.

Senior leaders of the 101st fully understood the sting of battle. For us, these battle-hardened warriors were determined that their subordinates would be fully trained and mentally prepared for anything that would come their way. The junior officers and enlisted were too young to have served in those wars. That did not stop us from benefiting from and learning from the climate the old guard created.

The Army was still dealing with problems from Korea when it converted to peacetime downsizing. It was not uncommon for a former captain to now be wearing the rank of staff sergeant. The same thing happened after World War II when lieutenant colonels were reverted to mid-level sergeants. It was very demeaning to the demoted soldier, but it happened en masse. On the downside for the rest of us, the result was the jamming up of the entire promotion system. Getting to junior sergeant could be done within three or four years, but after that, promotions to mid-level NCO ranks were almost impossible. On the upside, we had very skilled and combat-hardened leadership in the senior NCO ranks that had not suffered infiltration.

At the time, there was nothing I could do about officers who were reduced in rank who were clogging up the mid-level NCO ranks any more than I could about the failed promotion system that I had witnessed while in the Berlin Brigade. Not being able to influence the solution did not mean I forgot about the negative impact, but my involvement in a resolution would have to wait for another two decades.

One of the first actions after arriving at Fort Campbell was a trip to the consolidated equipment facility to receive all of my combat gear. Unlike supply operations in non-airborne commands, we required a lot of specialized equipment that was unfamiliar to me. One of our unit's senior NCOs was accompanying us on the initial issue. He had an aggressive attitude, plenty of jumps behind him, and enough stripes on his arms to be used on any supply clerk who tried to short-change us or issue defective equipment. Right there at the issue center, our NCO would make us put on all the gear, followed by his own thorough inspection.

If the equipment did not fit or were the least bit defective, he would force the supply staff to make it right. This served four purposes. First, it was the wrong time to find out your equipment was not fully functional *after* you jumped out of an airplane. Second, we only needed confidence in ourselves to jump out of that plane because we knew the equipment would work. Third, we saw our NCOs cared for and would use their stripes when necessary to protect us. Fourth, the supply staff preferred to get it right the first time rather than face confrontation with our NCO.

Reporting in as a non-airborne qualified infantryman came with the title of "leg." That title stayed until the completion of jump school (parachute training). Until then, we were assigned to the "ghost platoon." We had to eat at a different time than the rest of the company and live in a separate platoon bay. We had to earn the honor of intermingling with the airborne soldiers, and the rite of passage was graduation from jump school. We received no harassment or hazing from the airborne-qualified troops. Having no tolerance for such behavior, the NCOs would have immediately stopped any degrading behavior. The NCOs knew that those of us who graduated from jump school would soon be in company ranks. They saw to it that no seeds of discontent within the ranks would be sown because of inflated egos. "Rock Solid" always maintained itself as highly disciplined. So strong were the discipline and professionalism of that command that we seemed to recognize them with all five senses, including smell. Soldiers adjust to the professionalism, or lack of professionalism, to which they are exposed. In "Rock Solid," the professionalism of the soldiers was a direct result of the environment built and maintained by their NCOs and officers.

The ghost platoon lasted about a week before we temporarily moved over to jump school. While in ghost platoon, we went through various types of physical fitness training all day long. From the commander to the most junior NCOs, they wanted us to succeed and prepare us. They knew the physical demands we would soon endure. They gave us an edge that we put to good use.

Physically ready, once in jump school, we could focus on the technical aspects. First came the parachute ladder falls, which were done at different heights. Once we had mastered the techniques, we were ready for a thirty-four-foot tower. From that tower, we learned how to jump out of an aircraft, go down a cable, and stop at the end of the cable. Then came the three-day task of learning how to rig a parachute. Rigging consisted of putting the chute on and making sure everything worked properly so that when we jumped, we would have a safe descent to the ground. It is always comforting to know the chute and the jumper will come down together. The training was so thorough that we had no serious mishaps. A twisted ankle, if one did occur, did not qualify as a serious mishap. We learned that "pain is just weakness leaving the body."

After two weeks of preparation, we were ready to jump out of an aircraft. We were fortunate, as the availability of aircraft allowed us to complete our five jumps in a couple of days, rather than an entire week. Getting the jumps done in quick succession was good. The first jump was great because you really don't know what it's like. You just go up and jump out of the airplane and come down, reach the ground, get up, and turn in your parachute. The second one is the worst jump you'll ever make in your life because you understand what goes on exiting the aircraft. After that, you continue to progress until there is nothing to it. You get to the point where you have one hundred percent confidence in the equipment working. You know, too, how to exit the plane without causing any damage. You have confidence in yourself and everyone around you.

When we made our last jump, Westmoreland came out to pin airborne badge, or "jump wings," on us. He handed each of us fifty-five dollars. We thought that jump pay was the greatest thing in the world. We went back to the barracks where the company commander and NCOs shared their pride in us. With jump school completed, we became full-fledged members of the command. Now we could eat with the rest of the troops in the dining facility. However, we were not quite in the clear. Anyone with just five jumps is a "cherry-jumper." This term was to separate airborne soldiers from those who just went through the course for the sake of getting their wings and never exited a plane in flight again. I could not wait to get rid of that title. Two weeks later, during a command exercise, my sixth jump came.

During off-duty hours, soldiers were left alone. Living in platoon bay areas denied a lot of privacy, they did not need more senior soldiers invading what was left. Discipline in the barracks was not a problem. The first sergeant was single and always one door-knock away. The same went for unmarried squad leaders.

Unlike most of the soldiers, for me, going to the clubs and partying was more of an exception than a normal activity. One trip out taught me that not all social environments are friendly. The aftermath taught me something special. Unit cohesiveness means taking care of each other one hundred percent of the time. We did not fight in the barracks, but if someone attacked us outside of that setting, that was a different story. One night a friend of mine, Jerry Grandstaff, and I went to an off-base bar. There were a couple of nice-looking ladies and Jerry began talking to one of them. While I was in the restroom the lady's boyfriend arrived. By the time I returned he was beating on Jerry. I stepped in to stop the fight and get Jerry out of the bar. Suddenly the fight was fifteen civilians against us two soldiers. After thoroughly working us over, the civilians had a good laugh as they threw us out of the bar. We got a taxi and rode back to Fort Campbell.

When I walked into the company area, Staff Sergeant Kincaid, serving as the charge of quarters (CQ), said to me, "God dang, Specialist Gates, what in the hell happened to you?" I told him, and he went to the barracks, got about ten NCOs, and took us back to the bar. Kincaid's reaction force was still slightly outnumbered, but after a few airborne combative moves the imbalance of sides quickly turned to our favor. After beating the crap out of those civilians, Kincaid turned to the bar owner and made it clear that if he or anybody ever again struck an airborne soldier the NCOs were coming back and would close the bar.

The important lesson here was that we took care of each other. We did not start that fight. If we had, or if it had been one-on-one, that would have been different. Kincaid would have simply told us we should have known better. Being attacked by fifteen against two was something our NCOs were not going to ignore.

That was an example of why the battle groups had their own clubs on base as a place to relax and consume alcohol. The reason was to avoid being picked up by the police, getting traffic tickets, and to avoid getting into trouble by fighting with the local citizenry. Soldiers from other battle groups were allowed in only on invitation. All the girls from around the countryside used to come to those on-post clubs. If you wanted to see a girl, all you had to do was go to the NCO club or the young-enlisted club. If someone had too much to drink, he had fellow soldiers right there to walk him back to the nearby barracks.

Being a soldier in "Rock Solid" was made a lot easier and more rewarding by our two outstanding company commanders. I came in on the backside of First Lieutenant Bill Carpenter's tenure as company commander. About six months after my arrival, he would finish his tour of duty and moved on to a new assignment. Carpenter had a remarkable presence and demeanor, and his massive size was all muscle. Born in Pennsylvania in 1937, he was eight years old when his father was killed fighting the Germans in World War II. In 1958, Carpenter played split end as a member of West Point's undefeated football team. Legendary coach Earl H. "Red" Blaik identified Carpenter as "the greatest end I ever coached at West Point." The next year he was team captain. Twenty years later Carpenter was inducted into the College Football Hall of Fame.

Carpenter could have left the military after his contracted obligation was completed. Perhaps because of his father's sacrifice, but certainly because of his own character, Carpenter embraced a bigger obligation. That was to us and his country. He led by example. His leadership style fully reflected lessons learned from Coach Blaik. Like the football player he was, Carpenter knew how to move the field. Even though he selected a future that denied him a great legacy as a professional football player, another great legacy awaited him. I had no idea when Carpenter finished his command that we would meet again, under very different circumstances, thousands of miles away.

Most people would shy away from having to step into a command just completed by Carpenter. Taking the command to a higher level would seem impossible to most. Not for Captain Richard J. Cater. He was among the finest company commanders the Army ever produced. All leaders, Cater believed in tough, realistic training while enforcing all standards. He did everything possible to get people to move from one point to another. A firm believer in being both tactically and technically proficient, Cater was always creating an environment that enhanced our abilities and capabilities.

Cater didn't try to micromanage. During a platoon operation, he didn't come out and say, "Hey, guys, your fighting position is not done correctly." That was not the way he did it. It might come down to that, but to the platoon sergeant and the lieutenant . . . not in front of the troops. One of his techniques was after the Saturday morning inspections, he would stand on top of the steps going out of the barracks while we were out in formation, and he would ask questions like, "What is the weight of the M14 rifle? What is the purpose of the band around the end of the 106 recoilless rifle? What is the maximum effective range of the 81mm mortar? What are the types of ammunition? What are the parts of a call-for-fire mission?"

Cater knew all these answers from his own studying and experience. He pushed his lieutenants, sergeants, and junior enlisted during duty hours. During off-duty hours he pushed himself into the books and manuals. To keep up with him, in those platoon barracks and break areas, working with each other and our sergeants, we studied as hard as we could. Our NCOs trained us over the entire spectrum of an infantry company. Everybody in that company had to know how to shoot mortars, call in artillery and mortar fire and any other fire support we might need. Everybody had to learn how to shoot and maintain the 106mm recoilless rifle. We also had to know fire direction command procedures.

Like parts of a military-issue rifle or weapons system, we were trained to be interchangeable components. Extending beyond interchangeable weapons parts, soldiers were not a piece of cast metal capable of doing only one thing. Cater established an environment where we were trained to be flexible and use the broad skill sets we learned to fill other positions in the company. This included our knowing how to assume NCO and officer responsibilities in the event leadership was taken out. One way to assess a unit's worth is watching how it operates when its command element is gone. We would not have liked the thought of it, but we knew if Cater, his officers, and his senior NCOs fell in combat, his command would remain mission effective.

This kind of leadership was rather typical of the 101st at that time, but I also think we had a very special captain. It did not bother him to give people responsibility. One time we had an Armed Forces Day coming up, and the different battle groups were responsible for various static displays to entertain dignitaries and civilians, as they were waiting for our troops to parachute into the drop zone. Ours was to put up a display of equipment and various weapons. Showing his confidence in me, Cater simply said, "Sergeant Gates, this is your responsibility." He left all the details and coordination completely to me. He trusted my judgment and never said anything else until Armed Forces Day when he walked through the tent. He was satisfied with the result, and I was satisfied with his confidence in me.

When someone crossed the line and required discipline, Cater did not keep them guessing. In front of his desk, he had a brass platform marked with a set of boot prints that could only mean the position of attention. When someone reported for punishment he would say, "Mount the platform." He was not afraid of looking someone in the eye and enforcing the standards. At the same time, when the punishment was over, business returned to normal.

We had company-level dining facilities back then. There are a lot of positive things to be said about having consolidated mess facilities that replaced unit messing operations. More senior NCO management and a consolidated dining facility staff generally results in more and better food. However, something was lost in company troops not eating together. In the company mess halls, especially for breakfast, the officers and NCOs would eat in a separate area. That was "Commander's Call" where the day's work would be planned out, pending issues discussed, and coordination of effort resolved in a more relaxed environment.

Occasionally for dinner on Sundays, Cater would come into the dining facility. Sometimes he would join us in the NCO area, but usually Sundays were reserved for his time to eat with the young troops. Every time we went on a road march, Cater always wanted my squad to be the point, which meant our platoon was always in the lead. We thought that was appropriate. Like the lead dog pulling a snow sled, we had a change of scenery and knew the view everyone behind us had to endure. We were also proud that we were setting the pace, and we always did our best to ratchet it up as much as possible.

In the field, Cater and the platoon sergeants always enforced ten meter separation between soldiers, even if we were on the rifle range, in the chow line, or sitting down to eat. No automatic weapon or grenade was going to take more than one of us out. Also, it was easier to get away from a tossed grenade if soldiers had clear areas to jump into and were not bouncing against each other. That was drilled into me so hard that I later demanded it of my troops in Vietnam. In Vietnam, the enemy did throw grenades at us. Fortunately, I never had anyone killed by one.

Carpenter and Cater were outstanding officers, and in our platoon they were supported by one of the finest NCOs I ever knew. Sergeant First

Class Victor G. Franco was the epitome of a noncommissioned officer. He ensured we were trained and ready to execute any mission in the platoon. Cater set and enforced the standard. Franco made it happen.

Franco had a special skill of developing the NCOs, and he did the same thing for the young soldiers. He would ensure that no one in the platoon was referred to as anything but their rank and name. Derogatory name calling, even in jest, did not occur in Franco's platoon. The lesson he instilled in everyone was that every soldier is of equal importance. Officers and sergeants were treated with respect and Franco made sure that standard applied to the junior enlisted. If one of his soldiers had a pay problem, Franco did not just send him to the orderly room or to the finance office. Either he, or one of his subordinate sergeants, went with them.

I came under his supervision as a specialist just after completing jump school. One of the very first things he said to me was, "You ain't got no business being a specialist; you need to be a sergeant, and we're going to make you one." He first made me a fire-team leader, in preparation for squad leader. I thought, "Man, that makes me feel good."

Advancement to sergeant required going through the battle group promotion board approval process. First was the physical fitness test. Our first sergeant was not required to have his own board, but he did. No one in his command was going up to the Battle Group Board unless the first sergeant was satisfied we could do well. He was not looking out for his reputation. That was already well established throughout the command. The first sergeant was looking out for us. Franco prepared me for what to expect and provided the study materials. My implied task was not to disappoint either him or the first sergeant.

Long before the term "hip-pocket training" became an Army standard, Franco required all NCOs to have training cards in their pockets. He did not believe in dead time, instead he used those moments as targets of opportunity for impromptu training. Soldiers sitting around with nothing to do have time to get disgruntled. We did not have that problem. Franco did not dictate the subject of the impromptu training; it was our job to know the weaknesses we needed to address within our own squads.

In the legacy of a good NCO, Franco knew how to scrounge. Somehow, he got enough maps so that every soldier in that platoon had his own personal map of the maneuver area. Land navigation was his pet subject. Every time we would stop, Franco would ask the soldiers, "Where are you on the map?" If the soldier did not know, his sergeant had a problem. Those young soldiers learned how to navigate the land even without a compass, looking at the different terrain features. He did not require us to identify a ten-digit grid coordinate of our precise location, but we did have to be within a couple-hundred meters.

Training was the most important thing to him. Not only did we do tactical parachute operations, but we also did a lot of fire and maneuver. We did not have as much ammunition as we wanted, but nobody ever complained about not having live ammunition. If we did not have it, we either fired blanks, or got up there and, as strange as it sounds, just said "bang, bang, bang." Even in the garrison area without ammunition, we would do rifle drills and review the basics of firearms marksmanship. Ours was a professional environment that was tough to beat.

I became convinced that becoming a career soldier required maximization of one's own capabilities. To do that you had to earn it. Two of my best friends went to Officer Candidate School (OCS) and became lieutenants. I wanted to stay in the ranks and advance through the sergeant structure. With Franco's backing, I was sent to the 3rd Army NCO Academy. It was a different world then. They taught us to be squad leaders and platoon sergeants. We did not worry about spit-shining floors and polishing pipes; instead, the major emphasis at the 3rd Army NCO Academy was training. This included learning how to be professional during ceremonies and how to teach people to be precise at doing a left-face or a right-face or marching correctly in company formation. Platoon level tactical training went into great detail about proper fighting positions. As always, land navigation was a big thing.

Being that the 3rd Army NCO Academy was at Fort Jackson, it shared the post with the Women's Army Corps (WAC). They were great soldiers. At Fort Jackson they had their own club, and my classmates were over there every evening. They could not understand why I always opted to stay in my room studying. I understood I was in a class with very senior NCOs who knew more about the military than I did. The only way I could bridge that gap was to study as hard as possible. True, I did not achieve their fun memories of hanging out at the WAC's club. But, instead, when the course finished I walked away with being the [LB1] [CM2] distinguished honor graduate of the class.

Sergeant Franco was always encouraging us to go to any school available. Pushed by him and my own desire to learn, I went to jumpmaster school, followed by learning how to pack a parachute, at a difficult Air Transportability course. Training included determining how much could be put on a particular aircraft, where vehicles and other equipment would fit, and how to balance the load. It required a lot of calculating. Because computers and hand-held devices were still a decade away, everything was done with pen, pencil, and paper. We trained to do the job, but ultimately when the time came for action, the Air Force loadmaster did all the calculations. Our training allowed us to understand what he was doing and to know how to assist him in every way possible. Without the training, we would have been just another burden that he needed to overcome.

Recondo (reconnaissance commando) school was probably the toughest two-week course Fort Campbell had to offer. Of the thirty who entered with me, only twelve of us passed. Recondo was patrol training using a concept that was being mastered by then-Lieutenant Colonel Henry E. Emerson, who would later call it "checker-boarding." Platoons were assigned areas of responsibility, and once there, they would break down into squads and fire teams to find the enemy. We would go out, make a circle, come back, or go out to an intersection and come back. Once detected, the enemy would be held to a fixed position. Once the enemy was fixed, then the platoon piled on, especially from the enemy's flank. If the enemy was too strong for a platoon, then the company piled on. No full-scale attack was allowed until we had overwhelming superiority. Heavy casualties on our side meant someone failed. The Recondo strategy proved to be very effective in destroying the enemy while keeping our casualty rates low. It required a lot of training and coordination to master, that is why in peacetime we were training the way we would fight. Emerson always had a company in reserve that could quickly be moved in for the pile on. Later as a brigade and division commander in Vietnam, Emerson put this strategy to very good use.

After Recondo training, my next course was Ranger School. Being a junior sergeant in a course with senior NCOs and officers ranging from lieutenant to major meant paying close attention in class and studying every night. My coming back to Fort Campbell as the distinguished honor graduate this second time, as I had done at the NCO Academy, took massive effort and produced a major reward. A special ceremony was set up by Cater. The commanding general and brigade commander took charge of the ceremony and promoted me to staff sergeant. By studying hard and totally focus on completing the challenge of Ranger School, I was able to break through one level of the promotion barrier. If someone would later question why I got promoted to staff sergeant in five years while they had to wait twelve, I could quickly reply, "There's your answer. You waited. I worked."

Sergeant Franco was not done with me yet. After the promotion, he gave me more guidance. He said, "I'll tell you something, Sergeant Gates, what is that test you take when you first come in the Army? You got a 95 [score]. You are going to retake that test because you need 110 to 115. With your capabilities, you're going to retake it and probably will score much better."

Franco was right about the entry test I took immediately upon arrival at the reception station and before basic training. From not getting much sleep and all the travel and in-processing, everyone scores lower than normal. Scoring 100 reflects average intelligence. I was five points below that. Since that first test I had completed my General Educational Development certificate in Berlin, took numerous military courses, and had been constantly studying. Franco recognized this and knew my current score would harm my advancement in the future. While I did not, he saw the bigger, long-term picture. Complying with his orders, I retook the test. The score came back as 125. I would not have done it if not for Franco; that man always pushed me to be better. It was never beyond what I could handle, and he never put on me a bigger load than I could carry at any one time. As soon as I finished one accomplishment, however, he had another challenge waiting. It was not just me that Franco was developing. Fellow sergeants and squad leaders Tino Barajas and Bill Skinner came to the platoon with lots of skills and capabilities. Franco made them better, too. When Sergeant Franco moved on, Bill Skinner moved up and became an excellent platoon sergeant. From squad to corps, no command in the military should be dependent on one person to lead it. Within our company, Lieutenant Carpenter, Captain Cater, and Sergeant Franco filled that expectation.

CHAPTER 4
Impact of Great Leaders

One of the first events that made us realize the 1960s was going to be its own decade, and not a continuation of the 1950s, was the US Supreme Court's decision that the University of Mississippi's (Ole Miss) racial segregation against James H. Meredith violated the 14th Amendment. The issue had already been decided by the Supreme Court in 1954 with Brown v. Board of Education.

Some Southern universities had already started complying, although in small numbers. Ole Miss refused Meredith admission despite his qualifications and his court-proven Constitutional right to attend. Governor Ross R. Barnett was against the court decision and was going to use his state resources to defy the Supreme Court. Five years earlier, Arkansas Governor Orval E. Faubus activated his state's National Guard to block compliance with the Brown ruling. Then-President Eisenhower brought in the 101st Airborne Division to force compliance in what became known as the Little Rock School Integration Crisis. Not learning from this, in 1962, Governor Barnett tried to play the same game with President John F. Kennedy.

The result was identical. I was serving as CQ when the call came in from division headquarters ordering us to prepare to move. My next step was to call the company commander. In three hours, we were on an aircraft being flown to Memphis Naval Air Station. From there, we were trucked eighty miles to Oxford, Mississippi. Our mission initially was to set up roadblocks and checkpoints. Military police were soon brought in for civilian crowd control. For a week, we stayed in Oxford, within quick reaction time to the Ole Miss campus. After that, we were withdrawn back to Memphis Naval Air Station. Governor Barnett had no misunderstanding as to how fast we could get back to campus. The convoys to and from Oxford proved that. He also understood that, like President Eisenhower before him, President Kennedy was not a person to be underestimated. Courtesy of Governor Barnett's bigotry and arrogance, the entire nation got the same message.

President Kennedy had a unique ability to communicate with the people of the United States, both military and civilian. The charisma he had and the way he presented himself as President meant a lot in the military. Having been only seventeen years out of uniform himself, Kennedy still had a lot of military blood in him. Not for political reasons, he was always visiting military installations. We liked to think the 101st was special to him for having backed him up during two national crises. He came to Fort Campbell to visit everyone. The generals gave him their briefings, but we could tell he was there to visit with everyone, sergeants, lieutenants, and captains.

Because of the fact that he and his wife were so young, they really represented a new era. They also set a new standard for the positions of President of the United States and First Lady. It was like watching a movie, the way they carried themselves and the way they would speak and act in public. President Kennedy was not afraid to make a joke at his own expense. Because of Mrs. Kennedy's popularity when they traveled to Europe, Kennedy began his first speech in France by saying, "I do not think it altogether inappropriate to introduce myself to this audience. I am the man who accompanied Jacqueline Kennedy to Paris, and I have enjoyed it."

In the aftermath of Governor Barnett's ill-fated power play, I picked up another responsibility. Captain Cater knew the Berlin Brigade conducted extensive training in riot duty. Looking to potential future needs and recognizing the growing tension in the South, Cater decided we needed skill enhancement. Using the principle of knowing and exercising immediately available resources, he did not look far for his lead trainer.

Like the drop zone display, he simply assigned me this mission, knowing it would be accomplished to his satisfaction. For four months, I was training echelon lefts and rights, forward maneuvers, isolating rioters, and when to fix bayonets—all on order. Taking his place in the training formations and accepting instruction from me was Captain Cater. Had we deployed for a real riot, Cater would have been in the command-and-control position directly behind the line. To do it right, he wanted hands-on training to assess the capabilities and limitations of his soldiers.

During those early years at Fort Campbell, the ultimate opponents we were training for were the Soviet Union and the Warsaw Pact. Khrushchev was doing his saber rattling and declaring to the West, "We will bury you." As far as we were concerned, our "rendezvous with destiny" was to defeat the Soviets in Europe.

The flash point came when US spy planes detected Soviet missile systems being constructed in Cuba, ninety miles from Florida. Our entire division was restricted to base. Hearing President Kennedy's speeches on the radio, we knew we were going somewhere. In preparation, we started on the firing range to confirm the battle site zeros on our rifles, followed by weapons qualifications. Then came fire and maneuver training with live ammunition. This included individual weapons, crew-served weapons, and grenades. As we were doing this training, we observed the very quick buildup of about a hundred planes at Campbell Army Airfield.

Each battle group rotated between first-, second-, or third-ready alert status. First and second were restricted to barracks with no outside communications. Telephones were disconnected, and outside pay phones were off limits. Nobody could go home even if they were married. Trips to the post exchange were in small groups, with our leaders knowing our exact locations. Medical appointments were allowed, without deviations.

These trips outside the barracks during alert status came with the mandate to keep from talking about what was happening. The Army was several years away from developing operations security programs, which were created to protect the release of sensitive information. Back then, we did not need it. Loose tongues resulted in a trip to Captain Cater's office and a stand on that brass plate in front of his desk. If found guilty, coming off that plate would have had the impact of discipline. It would also result in a loss of trust from our teams and the confidence of our seniors. Those were prices no one wanted to pay. It was not just fear that kept soldiers from talking. Everyone knew saying nothing was the right thing to do.

We still did not know exactly where we were expected to deploy. It could have been to Germany to stop a potential Soviet advance into West Germany. All questions ended when we began pre-jump training with what we called water wings. These were nothing more than flotation devices for when we hit the water. The flotation device training was because of the hard lesson learned at Normandy: not all jumpers land on solid ground. No longer a secret that Cuba was our destination, Fort Campbell was our staging area. We were to be flown to Florida, refuel, and head straight to Cuba. The 101st was ready to go.

While still maintaining quick deployment capabilities, we started doing counterinsurgency training. Originally, this was called "counter-guerrilla warfare," but that was determined to be politically incorrect. We had already trained extensively at Fort Campbell and knew the routine. To remove us from terrain already known, we rotated in and out of Natchez Trace Park in Tennessee for training.

Someone must have studied the invasion of Cuba during the Spanish-American War. We commenced intense training in the numerous different geographical regions that could be found in different parts of Cuba. After the Natchez Trace, we went down to Florida and trained in the swamps. The company commander and the platoon sergeants had live .45–caliber ammunition; not to shoot people, but to kill snakes. Of all the snakes in the world, water moccasins have the worst attitude and are very territorial. Captain Cater got his kill while we were maneuvering in water up to our chests. Then came the Monongahela National Forest in the mountains of West Virginia. Even though we ultimately did not deploy to Cuba, this training would later prove to be very valuable. The mountains and the thickly vegetated overgrowth of this 900,000-acre federally owned forest were very similar to what we would soon see in central areas of Vietnam.

We lived out of our rucksacks. Those sacks were our homes. In them were five days of C-rations, a poncho, and a blanket. C-rations were the predecessor to meals ready to eat, but with food packaged in cans rather than plastic pouches. Tents were left behind at Fort Campbell. Even camouflaged tents were easily detectable. Entry and exit points were limited. Sleeping on the ground among the vegetation allowed for more concealment, plus the soldier had 360-degree tactical maneuverability if awakened. Our goal was to create a minimal footprint and go five days without being resupplied. If the ground was wet, we had a poncho to lie on. If it rained, we had a poncho to wear. We trained right. Soon, in Vietnam, we would realize that not all commands were trained the same way. We found that many troops did not know how to live out of a rucksack, and that would prove to be one of the many problems they would face.

The Monongahela Valley is famous for its whiskey, dating all the way back to the early 1800s. Taxes and production regulations drove the small distillers into an illegal profession. Unless they had a good place to hide, their stills had to be worked by moonlight to prevent law enforcement from seeing the smoke rising out of the trees, hence the term "moonshiners." My squad walked into the camp of a couple of old moonshiners making liquor in the woods. They had a smart operation. The Monongahela overgrowth and nearly unforgiving terrain were ideal for daytime whiskey brewing. They were deep in the forest and unlikely to be caught. They weren't operating on their own land, so if they had to quickly escape, there was no tracing the still back to them. Fortunately, we were in uniform with our tactical gear. They did not fire a shot. We had more guns than they did, but they had live ammunition. We assured them that we were not revenuers, politely declined their offer for a drink, and moved on. Just because they were running an illegal still did not make them any less patriotic and appreciative of our armed forces. Accepting their offer was out of the question. We needed to keep our heads clear for training. There was another concern in the back of my mind. Having grown up in the North Carolina hills, I knew the dangers of consuming home-brewed whiskey. All that aside, it was kind of amazing to see those moonshiners operating a still in the depths of a national park.

November 22, 1963, started out as a good day. In the afternoon, the Fort Campbell football stadium was packed because our team was playing against the 82d Airborne Division. Spirits were up on both sides. Everybody stopped talking when the commanding general made the announcement, "President Kennedy has been assassinated." It seemed like forever that everyone just stood there. You could hear a pin drop in that entire stadium as eight to ten thousand people stood in silence. We felt like we had lost a friend. Not only the Commander-in-Chief of the United States of America, but there was just something about him that made us feel like he was part of us, and that he really cared about us. We sensed that he did not have that separation from the people possessed by most presidents. We just stood there for about five minutes. The commanding general made the decision not to cancel the football game. Our teams went on and played it, but no one enjoyed it because all we did was talk about President Kennedy, what we could do about it, who was responsible, and who we were going to get for this.

During my time at Fort Campbell, through hard work and the impact of professional leaders, I progressed from being a young, fledgling soldier into a professional. There is an old saying, "You don't know what you don't know." While in the Berlin Brigade, I thought I knew a lot about soldiering. After a year at Fort Campbell, I realized how much I hadn't known upon arrival. It also brought the realization that there must be a whole lot more knowledge waiting to be discovered.

Like that young boy determined to pass on to the next grade after the bicycle accident, I was almost always engulfed in a book, a manual, or squad paperwork. Being "all I could be," meant hard work and studying. General Matthew Ridgeway, the legendary Commander of World War II and Korea, had already coined the leadership phrase, "You push troops during the day and push paper at night." That message I got as well.

There are people who claim the NCO Corps at that time were not as great as it is today. Back then, it was unusual for an NCO to have a college degree. Many soldiers and a few sergeants did not have high school diplomas. Upon enlistment, neither did I. For job proficiency, knowledge of weapons, instilling discipline and pride in soldiers, and taking care of soldiers, the NCOs that I worked with in both Berlin and at Fort Campbell were among the best that I have ever known.

The same went for the officers. Like the NCOs, officer counterparts in civilian life were making several times more in pay. None of our leaders were in it for the money. They were in for the long haul because they were professional soldiers who took pride in what they were doing and were dedicated to defending their nation. Their examples and their mentoring afforded us the opportunity to learn from them and be ready to one day take their places when it was our generation's turn to lead. In garrison, the senior NCOs and officers were the last to go through the chow line. In the field, they were the first people to be fed when a mess kitchen was set up. The reversal of order was so they could get their meals, get to work, and start coordinating the day while the soldiers ate. The squad leaders stayed with their troops. This was common sense.

No command can be better than its leadership. In the early 1960s, the 101st Airborne had the best in both seasoned officers and NCOs. Most of them had served in the 187th Regimental Combat Team during the Korean War. Even when they rotated out to other bases around the world, many of the NCOs were able to eventually return to the 187th. That command was like their home, and they were proud to maintain it in the best condition possible.

Fewer in number, but certainly more than enough to make a positive difference, were the veterans of the 101st or the 82d Airborne Divisions who had literally jumped into World War II. During those combat jumps, they were subject to being shot at while still in the air or immediately after hitting the ground. If any soldier broke a leg or was seriously injured, he would be stabilized until medics transported him to the hospital. The combat veterans leading us knew the importance of continuing to aggressively engage with and destroy an enemy. American soldiers already at the Bastogne perimeter in 1944 had held up well and had no intention of surrendering. The arrival of the 101st gave them a much-needed boost in seasoned warriors, weapons, leadership, and fighting spirit. That leadership and fighting spirit stayed within the 101st long after the wars in Europe and Korea were over. The respect we had for the 82d Airborne Division was not going to be readily admitted, but rather quietly accepted. Our drop zones were hard clay, and theirs were sand. We looked for every reason possible to prove we were the best division in the Army. At the time, that was probably true, but if called into combat, having the 82d on our flank would have provided a comforting feeling. Through strict discipline, our rock-solid NCOs trained us to very high standards. They knew every little nitpick of all the weapon systems in the company, and they made us train on all of them.

The 101st was the first division to receive the Light Anti-tank Weapon (LAW). The LAW was a significant improvement over the bazooka in that it was easy to carry and only required one person for transport and firing. The tube had to be opened, extended until locked into place, armed for firing, aimed, and fired while ensuring the back-blast did not wipe out the people behind you. A proficient soldier could do all this and destroy a target in less than fifteen seconds. Our World War II veterans recalled the days when they preferred using captured, more lethal, German panzerfausts (tanks fists) over American bazookas. We would later find the only downside of the LAWs in Vietnam. The tube had to be destroyed after firing. If not, the enemy could reuse it as a mortar tube. Smashing or crushing the expended tube got the job done.

We were also the first to receive the new M16 rifles. In the book *A Soldiers' Load*, Army historian Brig. Gen. S.L.A. Marshall addressed in detail how much weight he believed should be carried by a soldier moving to contact and engaged in combat. Marshall's recommended limit was forty-five pounds. The M16 helped to reduce the weight load as it was much lighter than the M14. The bullet was 5.56mm as opposed to the 7.62mm of the M14. The 30–round "banana clip" magazine allowed a lot of rounds to be fired without reloading. Because the M16 round was smaller and fired at high velocity, the round tumbled when it hit flesh and basically turned into a meat grinder. All these facts were on the positive side. Then came the negative.

The first rifles out were not a step up from the M14s. The M16 round had less penetrating power and could bounce off, be absorbed, or be deflected by wooden or metal defenses that the M14 round would pass through. The arc of the round was greater than the M14, which required judging distance from the target to allow sight compensation. The M16 was more prone to malfunction when dirty, especially with the bolt not locking into the ready position. This did not always happen, but when it did, it would prevent the weapon from being fired. Training took care of the trajectory arc problem. Increased weapons maintenance helped with the malfunction issue. Eventually, a thumb-operated "forward assist" would be installed. That problem should have been resolved before we ever saw the weapon. We later found that the ammunition used in the prototype M16 used a fast-burning propellant and did not dirty the action; however, the government contract for ammunition specified ball powder that created carbon buildup in the action.

Soldiers in combat should not be used for research and development for the military-industrial complex. The middle of a firefight is the wrong time to discover we have been issued defective equipment. If industry had as much dedication as our NCOs did when we were being issued our jump gear, then American soldiers and marines would not have deployed to Vietnam with flawed primary weapons. This lesson carried with me through all the ranks.

In addition to training us, our NCOs worked with the company commander to train the new lieutenants. There were clear lines of responsibility. In the concepts of "Be, Know, Do," the commander was responsible for the "Be" and ultimately for the "Do." "Know" required both technical and tactical proficiency. At the platoon level, this was clearly NCO territory. The lieutenants had academic and officer development schooling. Building upon that knowledge, integrating the lieutenants into the platoons required cooperation from the officer, the NCOs, and the troops. It was a work in progress that varied on an individual basis. Officers who lacked arrogance passed through the initial learning phases very quickly. Those who came from the enlisted ranks already had the knowledge; they just needed to recognize and accept that they were not in the ranks anymore.

Much of this training had to be done on a catch-as-catch-can basis. Most of the second lieutenants were on some type of additional duty when we were in the garrison. We hardly ever saw them except at physical fitness training in the morning and on Saturday morning inspections. Because of this, the NCOs were responsible for individual and team development in the garrison's closed-in training areas. These NCOs were as tough as they were colorful. Building soldiers from the inside out, they shared their combat knowledge with both the young officers and the junior enlisted.

As a squad leader, I had all the regular duties, like CQ, guard, and staff duty. Even as a junior sergeant, I could not help but realize that one of the dumbest things we had to guard was the Davy Crockett Tactical Nuclear Weapon System. Whenever we went to the field, it was with us. The fact that it was not nuclear armed was irrelevant. The system was classified top secret. Its purpose was to have nuclear capability in close tactical defense. I have no idea what Pentagon genius came up with the idea of using nuclear weapons in close tactical defense. The yield and the contamination fallout would probably have killed us all. The word "defense" implies that we were fighting to maintain allied territory. Why would we ever want to nuke the civilian allies we were supposed to be protecting from communist invaders?

It was during these years that Samuel T. Williams, who served as a general in both World War II and Vietnam, made the comment, "Esprit de corps doesn't get you to lean forward in the foxhole, it gets you to jump out of the foxhole and charge into the face of the enemy." He was right. Esprit de corps brings together unit cohesiveness. We in the 101st served as a quick strike reaction force. Quick strike meant that we were going straight to war, into the offensive, and we would start shooting—if necessary before we landed on the ground.

All the different battle groups had their own types of esprit de corps and different signs and symbols representing their commands. These were targets of opportunity for inter-unit competitions and pranks. We had a rocket plaque on our parade field. Troops from other commands would come late at night and sabotage it. That was fair, as we would do the same to them. Our revenge on the 506th Parachute Infantry Regiment was the best: they had a statue of a bull on their field—we castrated it with a hammer.

Fifty miles west of Fort Campbell, just across the border from Kentucky and on the west side of the Tennessee River, is the Paris Landing State Park. Having a nice beach, pavilion, and swings, it was a beautiful area. There were not a lot of recreational vehicles back then. People would go out in regular cars and pick up trucks to set up tents for camping. It did have a hotel run by the park service.

On Easter Sunday, 1962, a friend of mine and I drove out there to visit the park and enjoy the beach. As we drove up to the pavilion, there were two young girls trying to make the Coke machine work. They were putting money in, but nothing came out. I went up there and asked if I could help. One girl said, "This doesn't seem to work." We kicked it a little bit and hit it, put the money in, but it still did not work. When I turned toward her and looked into her blue-gray eyes, I realized she was the most beautiful blond-haired girl I have ever seen in my life. Everything about her, her eyes, hair, and dazzling smile on her face, was amazing. She stole that ol' hard-core Airborne soldier's heart right then.

All of us went up to a little service station just outside the park area, got some soft drinks, and sat down to relax. Then we went walking through the park. She sat down on a swing, and I pushed her. I knew I needed to see this lady again. I asked her if I could visit her the next day. She told me she lived fifty miles away, which made it about 100 miles from Fort Campbell. I made that trip.

Margaret Wilson originally came from a small farm in Pontotoc, Mississippi. Her family moved to about five miles east of Milan and had bought a service station and small grocery store, same thing my family did when I was a kid. We had a connection. The first meeting with the parents went well. I put on my best behavior, and we did receive permission to go to a movie. Before leaving to return to the base that night, I kissed her goodnight. I did continue to make that 100-mile trip as often as I could—and this was before the days of Interstates; it was just a two-lane highway. I got a couple of speeding tickets on those trips.

Margaret had a great family. They were all good, hard-working people. I really fell in love with Miss Margaret the first day I met her. Two years later, in June 1964, we were married in Milan. My platoon sergeant at that time, Bill Skinner, was my best man at the wedding. Many of her relatives were there, and I got to be good friends with all of them. She had a couple of cousins who became buddies of mine. When I took her to meet my family, my mother and father fell in love with her. My sisters thought it was the greatest thing in the world for Margaret to be one of their sisters. They never did call her sister-in-law; they always called her sister . . . Sister Margaret.

We moved to Fort Campbell and stayed in a trailer. I hate trailers to this day. She knew exactly how to set the table. I did not know anything about that, whether the knife went on the right side or the left. Margaret knew that. Cooking was a different matter. About a week after we were married and living in that trailer, she cooked dinner. The biscuits were hard as stones. Not being able to eat them, I said, "Margaret, this is great. You did a great job cooking these rocks." She picked up one of those "rocks" and threw it at me (laughing), and said, "Bill Gates, if you don't like what I cook for you, don't eat the stuff."

To make ends meet and to have a little better family lifestyle, Margaret immediately went to work in Clarksville, Tennessee, selling shoes. We knew she was not going to make a lot of money, but she believed throughout her entire life that she would get a job and work to make sure that the family lifestyle was a little better than what my pay was able to give us. There were times we had to pawn stuff to get through the month, such as my boots. Soldiers simply did not get paid much. Miss Margaret was quite a lady.

In September 1965, we had our first child. Melissa was born in the hospital at Fort Campbell, Kentucky. Melissa went through some hardship. For two of her first five years, I was gone. In May 1966, I went to Vietnam; came back in 1967; then went back in 1969. Margaret raised Melissa almost by herself, living all the time in that trailer. Margaret was a hard worker and an immaculate housekeeper. She had the charisma of a lady. She was always expanding her knowledge by reading. Margaret read every book she could. I don't think there's a word in the dictionary, a type of flower, or a kind of tree she did not know. She could remember all those words and points of the books she read. Margaret was very well versed in American history. She didn't really care about international history.

CHAPTER 5
Baptism by Fire

American service members of all ranks are notorious for not taking the time to thoroughly study the environment into which they are to be deployed. The combat experience of our seniors was gained in World War II and Korea. Almost never has America gone into a war that will be fought the same way as the last one. Except for fighting the Japanese in the island-hopping campaign in the South Pacific, we had no actual battle experience in what Vietnam would provide. Even World War II experience had limitations, because once an island was freed from Japanese occupation, it stayed that way. We would have gained more knowledge from the American campaigns in the Philippines, sixty years earlier, than from any other wars fought by the American military.

Secretary of Defense Robert S. McNamara visited Vietnam following the Battle of Ia Drang in 1965, led on the American side by then Lieutenant Colonel Harold G. Moore, Jr. After his discussion with Moore and others, McNamara returned to Washington D.C. and in his out brief to President Lyndon B. Johnson, McNamara assessed that America could not win a war in Vietnam. Yet, we would be in this war for another ten years. McNamara's assessment would be hidden from the American public for decades, well after President Johnson died of natural causes and

McNamara was in his final years. Instead, the reason we as a nation were always given for the Vietnam War was to prevent communist aggression and stop the domino effect.

American military forces were in Vietnam a lot earlier than most people know. In 1959, under President Eisenhower, there were 760 American service members in the country. In 1963, under President Kennedy, the number rose to 163,000. In 1968, under President Johnson, the number was at the high point of 536,000. All totaled, 2.7 million American military members served in Vietnam. Final totals of allied countries were South Korea—320,000, Australia—61,000, Philippines—10,000, and New Zealand—3,800.

The 1st Brigade, 101st Airborne Division, deployed to Vietnam in 1965. Assigned to the central highlands, its mission was to serve as a quick reaction force (QRF) for American commands in the region. At that time, individuals rotated in and out of the country. Once the command entered Vietnam, it stayed there. In 1966, I commenced my first tour of duty in Vietnam.

Transport to Vietnam was by commercial aircraft. Treatment on the flight from the stewardesses (as they were called in those days) was outstanding. The food kept coming. Once in Saigon, we went into a tent city for about a day. Not all of us were going to the same place. The replacement center sorted us out and scheduled our transportation. Mine was by C–130 tactical aircraft from Saigon to the US Air Force base at Phan Rang. Since my earliest days in the 101st Airborne, I began developing an ever-increasing admiration for the C–130 pilots. Propeller-driven, that plane was one of the most used and maneuverable fixed-wing aircraft in the world. Even the British Royal Air Force was flying them. The pilots had mastered every imaginable feat of tactical maneuverability out of those planes.

Originally, only an operational airstrip was built at Phan Rang. Over time, the Air Force constructed buildings and made it into a base. Nearby was our base, which consisted only of tents. Soldiers snapped their waterproof ponchos together to make a decent living space. This sleeping arrangement was one of the few situations where we did not maintain a ten-meter distance apart. Two soldiers slept in each double poncho hootch. Four days of indoctrination training were followed by transport to our battalions. Again, by C–130, we were flown from Phan Rang into a small village in Dac To. Located with a small American Special Forces camp, the 1st Brigade was there as a QRF. We did not arrive in Vietnam with our weapons. Those were issued in the country. Our brigade task force at Dac To consisted of the brigade headquarters, an artillery battalion, a combat service support battalion, and three maneuver battalions: the 2/502d, 1/327th, and 3/327th. The 2/327th was still back at Fort Campbell. I was assigned to B Company, 2d Battalion, 502d Airborne Infantry. My job was to be an airborne infantry squad leader and lead my soldiers in combat, do the best I could to save their lives, and beat the hell out of the enemy who was trying to kill my soldiers.

Since everybody came in as a replacement, there were always more seasoned men in each unit. Some soldiers had already been there eight or ten months. There was no need for some person from a different organization to come out and say, "Here, I am an expert on this stuff." We had our own experts. My squad was basically an even split between draftees and volunteers. In most cases, members of both groups had volunteered to serve in Vietnam. Whether drafted or enlisted by choice, they knew they had a job to do. They were good soldiers who formed a great team.

As in Dac To, once again, we were living in two-man poncho shelters. This buddy system allowed us to look out for each other, and we were an instant fire team in times of emergency. Squad leaders did not reside in the same hooch as the assistant squad leader. If the hooch had been hit by an incoming mortar round, the squad's entire senior leadership would not have been killed. The same rule applied to other leaders. Trenches had been dug and fighting positions established so we could defend the base. Whatever we had established on any given day, we were going to improve upon tomorrow.

At the time I arrived, the easiest way to identify a freshly assigned soldier was the condition of his boots. We deployed in-country wearing our standard-issued leather boots. Out on those forward operating bases, those boots lasted about a month. Staying wet all the time, the leather rotted on soldiers' feet, resulting in blisters and all kinds of other problems. The top of the boot separated from the soles. We did not have duct tape to try to extend the life of those boots. It is sad to think that the field soldiers of the greatest army in the world in the 1960s were suffering from footwear problems as did Washington's soldiers at Valley Forge, except we were suffering from jungle rot instead of frostbite. Eventually, after I was in the country for about six months, we were issued jungle boots. Before my arrival in Vietnam, some front-line soldiers had served an entire tour wearing rotting leather. It should not have taken that long for our quartermaster system and our senior officers to identify and resolve this problem. This was a clear-cut violation of "Care of the Soldier" principles. Those in the rear should have been ashamed of themselves for scarfing up equipment desperately needed outside the secure perimeter. Instead, they were too busy pretending to be and dressing like warriors. There were clearly two different worlds in Vietnam: the rear-echelon world, where people did not even need to carry weapons, and the world where the guys were fighting minute by minute to stay alive.

That kind of nonsense irritated all of us doing the fighting. A lot of people say, "I served in Vietnam." They did—in nice plush hotels in downtown

Saigon, on a base they never left, and with Vietnamese government organizations. Very few times, if ever, did they go into the field to live and fight as true American warriors. In cities like Saigon, they did not even carry weapons. All firearms were locked in arms rooms. Even in cities like Phan Rang, the threat was so non-existent that we had to leave our weapons on installations. In the rear, the military police had weapons, but that was it.

Those people in the big cities and large American bases came home from the war because they had a whole bunch of people securing them and preventing the enemy from reaching their safe areas. The frontline infantry soldiers over there did not live in such luxury. They earned their combat awards under fire, living out of a rucksack for a year out in the field. Awards were not just something that was given to them.

We maintained our airborne status and received jump pay while deployed to Vietnam. The assumption was that at some time the brigade would be required to make an airborne insertion in support of a future operation. We did rehearsals for an airborne operation and conducted one administrative jump. In the case of the 101st, we were accustomed to landing on hard ground and in small trees at Fort Campbell drop zones. Compared to some of the places we jumped into, a rice paddy would have been like landing on a pillow.

Even if a combat jump was not necessary, it made sense to have airborne troops, whether the 82d or 101st divisions, ready for quick reaction. As proven in World War II, airborne troops could move quickly in hostile territory, and with knowledge that they would probably be surrounded. From there, they knew how to fight their way out or inflict major casualties on an enemy being held in place, while other allied forces fought their way in. The 101st's ability to do exactly this was well proven at Bastogne. In

Vietnam, it would be proven time and again, except we would not jump in.

Even today, maintaining combat jump capability is necessary. Airborne is much faster than using helicopters. An Airborne operation can drop a few thousand people on the ground anywhere in the world in less than twenty-four hours. There will probably never be enough helicopters to move that many soldiers at one time. If the operation is in a distant country, the drop or landing zone will be beyond the range of the helicopters.

Three days into my assignment, we received an intelligence report that the camp was going to come under siege. The 95th NVA Regiment had been detected conducting movement to contact on our compound and the Special Forces' base, as well as the co-located village. It was assessed, and later proven correct, that their mission after defeating us was to destroy a couple of other small camps located nearby. Our mission was to stop them by inflicting as many enemy casualties as possible.

We packed our gear, including five days of rations, and went out into the operational area. For about three or four days, we conducted platoon-level recondos. The platoons set up little camp perimeters. The squads would go out and do U and V-shaped formations, checkerboarding, and then come report. If we detected anything that looked like an enemy movement, or enemy activity, or some type of well-used trail, we would report that to the platoon. Just as we were reporting to the platoon leaders, they were keeping the company commanders informed, who then reported to the battalion commander. Through his PRC–25 radio, our platoon leader could communicate with both us and the company commander.

Whenever we were on patrol, even if stopping for just an hour, we would set up a defensive perimeter. We would know automatically what to do when put in a certain position. Two- or sometimes three-man fire teams were immediately assigned sectors of fire within squads. Then we made sure our machine guns were properly positioned. Within five minutes, that would all be done. We did not make range cards for these stops, but every person did their own range estimate. Ninety-nine percent of the time M60 machine guns were on a bipod that was already attached to the weapon. We did not dig foxholes, but we would build berm shields with what was available. The enemy could use the time we were at that location to prepare and execute an attack. It would have been very inhospitable of us not to welcome them into our camp without a proper reception.

In-depth prisoner interrogation was done in the rear by the G–2 division intelligence. The Viet Cong (VC) were more prone to talk than the NVA. We did try to conduct tactical interrogations. The impediment to this was having someone we trusted who spoke both Vietnamese and English. Except for the indigenous people of the central highlands called Montagnards, our trust in the Vietnamese assigned to us rarely existed. The second problem was language skills. The average interpreter we had could not speak English very well; subsequently, they were not very useful. The best interpreters we had were American soldiers who had completed language school before deploying. We had some in our unit. Most of the Special Forces could speak Vietnamese fluently. Certainly, G–2 section had personnel who were fluent.

On one mission, the squad was following a trail into known NVA territory. Using battle-proven Tactics, Techniques, and Procedures (TTPs), we were moving on both sides of the trail. Our point man detected an NVA position on a small hill to our front. Proper exercising of fire and maneuver saved us from a disaster as they were in an excellent ambush position. Any failures on their part would cost them their lives. The ambush had an excellent field of fire, but they did not maintain proper concealment. That, and their failure to exercise noise discipline, caused their discovery. They also did not pay attention to their entire surroundings. Army Rangers never would have made any of those mistakes.

We developed an "L-shape" assault formation. Once my fire team was on the flank, having already taken effective cover, the assistant squad leader's team engaged the NVA in a firefight. The enemy did not realize that the gunfight was a distraction until the assault team was on them. When the fight was over, the satisfaction of no one wounded or killed on our side was short-lived. One of the soldiers said to me, "Sergeant Gates, you're bleeding." My response was denial. "I'm not bleeding." The soldier pointed to the blood soaking up my lower shirt and repeated himself. He was right. Pulling up the shirt revealed a bullet wound that went across my stomach. The bullet did not lodge itself but made a cut as if done by a sharp knife.

The wound could not be left alone. It was beyond handling with field dressing. Without proper stitchwork the result would have been a bleeding out. Had we subsequently passed through waist-level water, the leeches would have formed a buffet line. A medevac was required. Temporary departure from the troops was an irritation, but they were in good hands with the assistant squad leader. As always, the doctors and nurses worked their magic. They also exercised their authority in denying my request for an immediate return to the company. No doubt, they had previously dealt with hundreds of wounded soldiers, forsaking common sense in favor of wanting to immediately return to their battle buddies. For five days, they ensured an infection did not set in. Once it cleared, I was on my way back to my command. When the proper time came, the unit medic removed the stitches.

Patrolling requirements did not take holidays. Christmas patrol earned me a reprimand when my squad returned to camp. We had passed through a small farming village. Being it was Christmas, my soldiers gave all their food rations, ponchos, and other life support items to the children. I did not know of their generosity until it was too late. We could not go back and recover the items. In accepting the reprimand, I felt proud of my soldiers for remembering Christmas and sharing gifts with those less fortunate. While being surrounded by war, they had not forgotten their family values.

There is a misconception that the central highlands consist of barren mountains. The Vietnam forests have triple-canopy foliage and streams come from everywhere. During the heavy rains, those streams turned into small rivers. A squad patrol could take four hours to travel just 1,000 meters. Dealing with enemy activity and booby traps extended that time much further. Some of our movements had to be cleared by a machete as we went.

On another mission, we were down in the low ground between some mountain hills and had a huge ridge line next to us. The battalion commander, positioned inside a helicopter flying over our position, spotted a possible trail about halfway up the ridge line. We did a reconnaissance up the ridge line and, sure enough, there was a well-used trail, and a blue communications cable appeared. Applying indications and warnings, we knew there was something more than just VC in the area. Our assessment was that the trail was a major network for the 95th NVA Regiment, an NVA regiment that generally consisted of between one thousand and twelve hundred men. Very soon, our flank guards warned us of an individual walking down the trail wearing a khaki uniform and a pith helmet, which was typical for the NVA. With all that foliage around, it was not hard to quickly camouflage the entire platoon. When he got next to our center, we opened fire. He turned out to be an NVA officer. We confiscated his pistol and the documents he was carrying to be turned over later to our battalion headquarters.

We started moving down that trail. It was my turn to be the point squad for the platoon. I was right behind my point man, Specialist Jesse James (actual name). Most of us have five senses. James had six. I do not know what it was about him, but he had a capability that allowed him to detect human beings in the area. Every time he would stop and tell me there was somebody in the area, he would turn out to be right.

A good infantryman wants to be able to pick out those things that are critical for any particular operation. If you see a tree cracked or a limb broken, that is a reason to stop, investigate, check it out, recount it, and see what it is. If you see a trail, it may be an animal trail, but at least you stopped, looked at it, and determined what it was that caused something to be different. It takes a lot longer than one patrol for that to be learned. If not learned, someone could get killed or wounded.

We were moving quietly. Coming around a bend in the trail, James was alerted to an individual up ahead manning an air defense gun mounted on a tripod. He was at the edge of a small clearing and waiting to shoot at passing helicopters. We would soon find out why he was posted there. Just as he realized our presence, we shot him. He went down from being hit but somehow got away. The best we could assess is that he went into a tunnel, which we could not find.

From there, we kept moving down the trail and came to an open area that would facilitate a helicopter. The platoon leader stopped us and called me back to where he was behind my squad. He was having a radio conversation with the company commander, as about three kilometers away, C Company of the 502d had come into heavy NVA contact. There was no leadership deficit in Charlie Company, as it was commanded by now Captain Bill Carpenter.

We got the mission to move to C Company and assist them because they had been surrounded for about three days. Our job was to break through the siege and help them. Moving three kilometers through that foliage was not going to happen quickly because we had to tactically deploy, keeping ten meters between all members, and not walk into a trap ourselves.

That NVA soldier on the air defense gun was our indication and warning that this attack on Carpenter's company was well planned and well thought out. The NVA soldier's escape gave him the opportunity to alert his seniors of our presence and that we were moving toward the company. The NVA was going to have troops ready to engage us. The most tactical thing they could do was go all in on destroying C Company before we arrived and then go all in on us. Apparently, that's what they tried to do.

There is a fighting philosophy of "hit 'em hard, hit 'em fast, and hit 'em with the one they don't expect." Doing exactly that, just when C Company was about to be overrun, Carpenter called an artillery strike on and near his own position. With the warning received from Carpenter, C Company had just enough time to make use of what cover they had available. NVA never saw it coming until American supporting fire ripped them apart. Artillery fragmentation is very unforgiving.

There were some injuries to C Company caused by the artillery. Because they had been able to take some cover and were in better physical shape than the NVA soldiers, the Americans were able to recover more quickly and resume the fight much faster. The siege was broken for the time being, but the fight was far from over.

In the beginning, C Company found itself caught in a fight with an entire NVA regiment on their terrain. When a command calls out "Broken Arrow," that means all American resources possible will be coming to the rescue. The NVA lost a lot of people, not only from our individual ability to hit them with mortars, small arms fire, and artillery fire, but also because we had helicopters with door gunners and, eventually, missiles on helicopters. That enemy soldier was stationed in a clearing with an air defense gun because of the NVA's anticipation of American helicopters joining the fight. What they could never have predicted was dealing with an American warrior who would call fire support onto his own position. That bold move had to have stunned the NVA commander and made them rethink their strategy.

With my squad still working point for the platoon, we continued in our movement to link up with C Company. Coming upon an open area, we stopped. An open area in a hot combat zone presents an excellent field of fire for the enemy. Securing it required the coordinated effort of the entire platoon. I walked back to the platoon leader and the platoon sergeant who were immediately following my squad. Why did it not happen to me, I don't know, but there was a sniper located down there. On my walk back, I was passing one of my soldiers who took a sniper shot right through the face, from one side of his jaw to the other. Even still, I often wonder why the sniper did not shoot me. I don't know, because I was a great target.

It seemed like forever to get this young man medically evacuated. The helicopter could not land in that clearing until it was secured. Right beside that clearing was a ridge line and during our clearing process we discovered it was occupied by NVA soldiers in a bunker. They were waiting for an expected air assault to arrive in support of C Company so it was fortunate our commanders did not take the bait as it would have been a catastrophe.

For all of us in the platoon, seeing one of our team members shot like that took something out of us. It also put something into us. We got angry at the NVA and more determined to kill some of them—to fix our bayonets on our rifles and to go after them. Everybody in the platoon was mad, fighting mad. If left uncontrolled, that would have been stupid for a variety of reasons in a combat situation. As professional soldiers, we knew we had to channel our anger and put it to proper use in engaging the enemy. Immediately regaining control of our emotions, we focused on our surroundings and the fight in which we were engaged. Not from a ladder, but from a rope hanging from the tree, we found where the sniper was hiding. Once we locked in on him, we took him out with small arms fire. Then we assaulted the bunker area. We moved around to the side. Flanking that bunker, we took it from the enemy and secured the open area. In the process, we captured a couple of NVA soldiers.

The first wise thing about controlling our anger was that we did not allow ourselves to become sloppy in exercising our fire and maneuver tactics. We had no casualties in taking that bunker. The second wise thing was not allowing our anger to motivate us into committing war crimes against those NVA soldiers. Once they surrendered, their well-being became our responsibility. Professional soldiers do not commit atrocities, including against prisoners of war or anyone who is not an active combatant. It took us about an hour to secure that area, and eventually, the Huey medevac helicopter did come in and pick up my soldier. By that time, the soldier was dead even though the platoon medic worked his butt off trying to save him.

We proceeded to work our way the remaining two kilometers to C Company, which took about two days, but we finally linked up. That NVA regiment was all around in that area. Moving through that type of terrain was very tough. Enemy contact made it worse. Together with Carpenter's troops, we pulled back as a single operation. When we were well clear of the area, the fight temporarily belonged to the Air Force. B-52s were on standby with bellies full of 500-pound bombs. That meant we had to get more than two kilometers away from the area targeted for carpet bombing. Those planes were a great asset, but they could not be very accurate. That was not their intent, their objective was to clear an entire area. Once we radioed back that we were clear, the pilots came in and delivered their payloads. This lasted for nearly half a day. Those 500-pound bombs produced ten-foot-deep craters. The incendiary mixture of petroleum fuel and a gelling agent within the napalm bombs turned the place into an inferno. Anyone caught in the area of saturation was annihilated.

We went back in to assess the enemy battle damage. The carpet bombing had been successful. Had Carpenter ordered a breakout from his position, his command would have been slaughtered because the NVA regiment was well situated. All around his perimeter, away from the area of intense fighting, the number of defense positions was amazing. The NVA soldiers had constructed trenches so well camouflaged that even after the bombing, we could hardly see them. They also had set up booby traps including ones with hand grenades and trip wires. They had dug rectangular holes in the ground with bamboo sharpened punji sticks angled downward from the walls of the holes. Soldiers would have stepped into the pits, and the sticks would give just enough to wedge against the soldier's leg inside the pit. The sticks would stick into the shin and calf, preventing soldiers from quickly pulling their legs out without making the injury worse. The sticks would have had to be either cut or dug out. If the sticks were poisoned, there was not a lot of time available, especially considering the firefight would still have been going on.

The sophistication of booby-traps, out of the most rudimentary of natural resources, was not a sudden invention of the North Vietnamese military. They had generations of internal conflicts to develop their warfighting skills. Their booby-traps had already been battle-proven against Western nations in warring with the French military. It was wise of C Company troops to stay together and fight as a team. If any of them had tried to break and escape for their own safety, they would have run straight into all those booby traps and enemy fire. It was also wise that, despite moving as fast as we could, our company moved with tactical caution. Both companies survived because everyone kept their heads and used the training tactics they learned at Fort Campbell and in Vietnam.

After completing our battlefield assessment, we were pulled back to the brigade base camp. Part of the brigade then pulled back to a place called Tuy Hòa on the beach. In our out-briefs, we were told that we had been very successful in stopping the 95th NVA. We may have been given an estimated body count. At my level of concern, I did not pay attention to what we were told. Body counts had more to do with politics to make us look good to the American people.

CHAPTER 6
Assessment and Reflections of the Vietnam War

With what Captain Carpenter's units had attacked them, other commands may well have broken. The same applies to what happened at Ia Drang when Colonel Moore's battalion was severely outnumbered by the enemy. It must be remembered that Carpenter was a protégé of Moore. They handled their commands the same way. That was not the case with one of the two battalions sent in to help extract Moore's mauled, but victorious, battalion. After Moore's battalion had departed, Lieutenant Colonels Robert McDade and Robert Tully, had to get their commands out by foot. McDade's battalion marched out in a single file with no security posted, while Tully's deployed tactical maneuvering all the way down to squad level. McDade's troops got hit hard; they would have been completely slaughtered had Tully's command not come to the rescue.

It is no surprise that Moore and Carpenter retired as highly respected lieutenant generals. Both understood what would happen to your organization if you did not fight in an offensive mode. Even on the defense, they knew how to conduct an offensive operation. Carpenter and Moore were always examining what they could do to stop the enemy from attaining a certain goal. In combat, the mindset must always be offensive, offensive, offensive, and offensive. Even inside a perimeter, that means finding the key to survival. In turn, there is always the thought of how to obtain and use the offensive tools necessary to fight and defeat the enemy. Both of those officers were totally committed to offensive operations, not defensive. In the base camps, their perimeters were well fortified. They had also identified all the grid coordinates outside their perimeters for calling in artillery in the event of an attack. They had war-gamed through counterattack procedures.

Once they made contact, they started maximizing the capability of all the systems they had available. Moore passed on to his subordinates the lesson of not receiving everything desired. That is always to be expected, especially in combat. The key is that he and Carpenter were able to maximize what they did receive, be it artillery, armor, helicopters, or fixed-wing aircraft. The lesson is that whimpering about what support is not available never helps to overcome any situation. This applies to both combat and peacetime operations.

Both officers enhanced their formal training with additional self-study. They knew how to do an immediate risk assessment, knowing close air support and artillery may result in losing some of their own fine troops. In the middle of the Ia Drang fight, a bomb dropped from an American aircraft flying at tree level landed inside Moore's perimeter. Moore and his Sergeant Major Basil L. Plumley took quick action and focused their troops back on their external fight. It is never desirable to lose anybody in a fight. Risking an entire company or battalion by avoiding actions that could endanger a few is even more undesirable.

Since Carpenter knew the close air support was coming in, he was able to get out the warning to his men. There has been talk about how Carpenter's organization suffered burns from his calling in the close air support. That is only part of the truth; he did it to save the command—and he succeeded in. That is what a commander does. I was part of the company that came to back up Carpenter. From the enemy bodies on the ground, I saw how close the fight had been before he called in the strike.

I knew Carpenter. He was well-versed in team operations, had been to Ranger School, was well-trained, and had been through hard courses at Fort Campbell. Although combat was new to him, being a company commander was not new. His training and character allowed him to quickly adapt. Carpenter did a good job during that battle and wherever else he went. When we pulled into his perimeter, none of the men of his command had anything but praise for his actions. The same happened with Moore's troops when their relief battalions arrived.

Carpenter and Moore always took time to develop their soldiers into warriors. They both knew the meaning of the maxim, "If you get caught in a fair fight, you did not come prepared." Both were among the best warriors to ever be found on the battlefield. It was not just because of their direct involvement with their soldiers, it was because they had created environments where outstanding subordinate officers and NCOs in their commands succeeded at doing their jobs. Good NCOs develop soldiers all the way from the most junior private. Another of Patton's philosophies were "loyalty works two ways—from the bottom to the top and from the top to the bottom." Carpenter and Moore never had a deficit of two-way loyalty in their commands. We always had it in our command as well.

There were not a lot of Americans killed in action in Carpenter's battle, but a lot of soldiers were wounded. The 95th NVA sustained much heavier casualties. The difference between American and NVA casualties was not just because of the infantry soldiers on the ground. Without all the coordinated and combined efforts of multiple Army and Air Force units, we never would have gotten to C Company in time. On the way, we would have been torn apart as well. Credit needs to be given to the warriors who came to our rescue. Just as they did it for us in that one battle, they did it all across Vietnam throughout the entire war.

Those Huey helicopter pilots and door gunners were outstanding. They would come in and do their best to strike hard at the enemy attacking our perimeters. The pilots would launch the rockets, and from each side of the aircraft, the door gunners would rip into the enemy from angles and with firepower not available to us on the ground. After making their strike, rather than flying away, they would come in and land. Working with ground troops, the flight crews would unload ammunition and supplies. Before leaving, they would load up their aircraft with the wounded requiring surgical attention. Tens of thousands of soldiers wounded in combat survived because those flight crews put their own lives at risk.

The pilots would set down on landing zones where no helicopter would expect to land. Even at night, when those guys should not have been flying, they were still providing us fire and transport support. Helicopters would get shot up and even taken down by enemy fire, yet the aviation warriors kept coming to our aid. Anyone who says a bad thing about Army aviation speaks solely from ignorance. Those guys really came through for us.

The pilots and the door gunners were just like us. Same with artillery. They would do everything possible to get rounds shot down range. Even if impossible, they'd still try to do it. Working closely with the pilots back at base camp were our own unit-level logistics sergeants. While we were fighting, they maintained close contact with our leaders. When we went out for thirty days, only five days of rations went with us. It was the supply sergeants who got additional rations and life support materials to us. If something needed was not in their supply inventories, they would use their networks to find it. No one can wheel and deal like a supply sergeant. We were too busy to ask and really did not care how they got it to us. We were just appreciative of what they were doing.

A special group of people deserving recognition was medics. Not only did they come in with medical supplies, but under fire, they would try to do everything possible to save a soldier. They would also fight right beside us. One replacement medic was a conscientious objector. That lasted about a week, then he willingly turned into a combat infantry soldier when we got into a tight spot.

We had Air Force members attached to our company serving as medics, engineers, and in other specialties. That's how it was in combat. We all helped each other and fought for each other. Everybody talks about the flag. In the middle of a firefight, you are not fighting for the flag. You fight for your buddy on the left and right so they can live through another night. They are doing the same thing for you. It does not matter what uniform is being worn; they're American warriors, and that's all that matters.

After our mission to block the 95th NVA was completed, there were no more major conflicts for the rest of the time I was there. That battle caused enough damage to the 95th NVA Regiment to make them ineffective at the combat command level. It took two more years of reconstitution until the 95th could launch an offensive operation in that area. In the meantime, they continued small-level hostile operations against us. We patrolled areas around Dac To and then went to a place called Con Thien. We moved from there to Tuy Hòa on the coast. There was a small civilian airport that the military had taken over for our heavy-lift Air Force planes. We helped secure the airport and conduct counterinsurgency operations along the coast and into the mountains to the west. We did most operations by air insertion and would stay there for about thirty days.

We always had to deal with the irregulars, the VC; nighttime was normally when the VC would conduct their operations. We had a couple of pieces of equipment for night maneuvering, but none of it was worth the effort to use. The equipment was a poor comparison to what American warriors have today. Night-vision Starlight periscopes were mostly it. What we mostly used to conduct night operations was common sense and our tactical ability to navigate through darkness.

How the VC would conduct their operations was obvious; they would use the availability of little trail networks that ran all through the vegetation. The VC did not try to slip through the jungle canopies. Once we learned this, we got smart and would ambush those trails. We would set up ambushes on any trail or trail network in the area of operations (AO). It was very highly successful. That's why there were curfews. If anyone was out at night in our AO, they would be considered an enemy combatant. This applied only to the countryside. Anyone moving at night would likely get shot because we usually had ambushes on those trails, and those ambushes included claymore mines. Once the claymores went off, we would open fire. We had already set up our range stakes, establishing the limits of left and right, as well as up and down. Staying within those parameters, we ensured we were getting maximum kill on the enemy and not shooting toward other platoon ambush locations.

It was not easy moving all day and then staying up at night, losing fifty to seventy-five percent of necessary sleep, even sometimes one hundred percent. We developed a smart technique. In the evening, we would set up a perimeter before we took out our C-rations and made some coffee. We did this on purpose. We made a little noise, cooked, smoked, and such. While some of us were doing this, other team members would conduct a reconnaissance of the area where we intended to set up the ambushes that night. When it got dark, we would completely vacate the perimeter and quietly move to the ambush sites. The VC would come over and start shooting at our empty camp. That is when we would open up and fire on them. Many times, they would be coming down the trail trying to hit the site we were supposed to be at and walking into our ambushes. Old American ingenuity came through over there.

We also used nature to our advantage at those ambush sites. After being positioned for an hour or so, provided we were quiet enough and did not make any noise, the birds started singing, and the animals started moving. When wildlife activity quiets down suddenly, something is happening, and someone is coming. That was our indication and warning that we were about to fight. All land navigation was done by a compass with a map. It was easy to figure out routes this way, even though some maps were not very accurate. I turned out to be the platoon navigator because of my experience in land navigation at Fort Campbell, going through Ranger School, and my on-ground experience. Our mindset was on alert because, at all times, the enemy could be just yards away.

Reminiscent of our own Native Americans, the Montagnards had their own villages all through our area of operation. Even on the mountain side, they raised rice. The paddies were terraced, allowing for downflowing water to irrigate the next terrace below. Sometimes, from a helicopter, we would see where they had cut down trees from the mountainside and planted rice. They were good people, favoring Americans over the Vietnamese. We came with our own rations and took nothing from them, unlike the South Vietnamese. The North Vietnamese often committed atrocities on the Montagnards. Many of the Montagnards were trained by our Special Forces and had indigenous forces well-trained from their communities throughout the central highlands. They were good soldiers, and they knew the area well and were very dedicated to us. We often would have one or two of them with us as scouts for the platoon. They were difficult to understand but very good, hard-working people.

The weather there was a problem. During the monsoon season, a period of about three months, we got an enormous amount of rain. It rained every day, seeming like twenty-four hours a day. We stayed wet the whole time. It still rained quite a bit during the remainder of the season. It was hot during the day. Humidity was terrible in the summer and during the monsoon season. But even in the summertime, when it was hot and humid during the daytime, we got cold at night simply because we couldn't dry off from all the rain.

Unless closely monitored, being in the field could become a medical nightmare. Our feet suffered from jungle rot because of bacteria, and all the water we were dealing with was a major issue. We had continuous problems with malaria and dysentery, and there were millions of mosquitoes requiring us to have to take daily malaria pills. Many soldiers contracted amoebic dysentery. Everything they ate went straight through them. There was no medicine to stop it. The only solution was to medevac to a hospital and stay in a doctor's care for a couple of weeks.

The leeches were a far bigger concern. Once on the body, those parasites would start sucking blood and eventually get their head under the skin and cause an infection. The best way to get rid of them was to take a cigarette and burn them. Insect repellant was not helpful as leeches seemed to love that stuff. Wearing leather boots that rotted away within a couple of months prevented leeches from attaching themselves to our feet. If we were not paying attention, leeches would crawl up our legs and attach themselves around our waists, which created the most problems. Failing to notice and remove the leeches in time resulted in serious medical concerns. Soldiers and their leaders did the best they could to keep the leech problem under control. However, movement to contact against the enemy and firefights often prevented frequent checking. One soldier did get seriously infected from a leach. His attempt to receive a Purple Heart for the infection was properly rejected. Leeches were not on our list of enemy combatants.

Without any incident, we occasionally encountered elephants. Coming across their footprints was much more common, in part because the NVA was using them to transport equipment. Knowing that elephants are intelligent and aggressive only when they feel threatened, we ensured the wild ones were given a wide berth.

Then, there were spitting cobras, pythons, and other snakes. One day, the platoon was moving up a trail network. We were in proper patrol formation and not clustered on the trail. I was at the rear of the first squad. Suddenly, we stopped, and I ran up to the front and asked what had happened. Probably about a foot in diameter, curled up in a coil right in the middle of the trail, was a snake. We were afraid of that snake since it was the first time we had seen one of that size. We ended up shooting it, although it probably would not have hurt us unless we had stuck our heads right next to him.

There are two types of venom from snakes: hemotoxic and neurotoxic. The snakes with the hemotoxins were pretty. You can survive their venom for a brief period. If you do not get aid for the venom, it affects your nervous system, and then you will have a problem. Plenty of them were there trying to get into the trees and bushes. If we sat down, they would ease down and maybe get on the soldier's back or rucksack and end up hidden. Those snakes with neurotoxin venom were dangerous. I had one soldier bitten by one. If we had not medevac'd him, it would have killed him. He was medevac'd all the way back to Fort Sam Houston, Texas. It did affect his nervous system, and he never came back to the unit.

Chapter 6

Looking back on history, when has anyone ever won what they call a counter-insurgency war? There's an example of the British in Borneo. They stopped the insurgency, but who was running the country as soon as it was over? The people who the British were fighting—members of the insurgency. That lesson escaped those in the highest levels of decision-making. Insurgents normally win the wars. They might not win the battles, but they will win the war. In wars like that, caution must be taken in how a nation gets into it. Our government in Washington never took any of this into consideration.

We had two fights. One was from the North Vietnamese. The other was from people who were tired of foreign intervention. The French had come in, owned the property, and treated the Vietnamese people as slaves. Further agitating the citizens of South Vietnam was the corrupt government. The villagers and farmers were trying to survive with what they could produce from the land. The national government in Saigon and the provincial governments were corporate raiding from the people.

The South Vietnamese had some outstanding, dedicated soldiers. They did some hard and intensive fighting. A lot of blood from the South Vietnamese Army was spilled fighting the NVA and the VC. The sad truth is that they were more dedicated to the future and welfare of their country and its citizens than were the politicians and self-serving bureaucrats running the country. The fact that our own government, and especially our State Department, continually kept backing the Saigon government added to our problems in the field. Even the 1963 CIA-backed coup against the Diem government just resulted in more corrupt officials moving into position.

Starting with the assessment that Secretary of Defense (SecDef) Robert McNamara had already given President Johnson that the United States and its allies could not win the war in Vietnam, several other problems began developing within the US Army. By the end of the decade, professionalism within the ranks would take a pounding. A perfect storm was in the making.

There were tactical plans in Vietnam, but no strategic plan had been developed. Telling us we were going in to help the South Vietnamese was not a strategy. That was not even a philosophy. At best, it was a rallying call that wore itself out. A war cannot be won without a strategy. Instead, Vietnam was piecemeal, always piecemeal. Warriors despise paying for the same real estate twice. In Vietnam, we had to pay for it time and again, often with our lives and limbs.

One of the worst mistakes ever made was President Johnson's refusal to activate the citizen army. The total Army is the Active Component, the Army Reserve and the Army National Guard. There would have been nothing wrong with mobilizing entire commands and individual soldiers in the Reserve or the National Guard. By making it clear he would not do that, President Johnson made the Reserve and National Guard a sanctuary for draft-eligible men to avoid active service. Draft dodgers had four options: go to college, go to Canada, go to prison, or join the Reserve or Guard. President Johnson's flawed logic was that it was better not to disrupt American society by removing the family breadwinners from their employment and homes. This was nonsense. Leaving the Reserve and Guard at their drill stations resulted in a major inflation of active-duty ranks serving in Vietnam. Young men who had dropped out of high school or just finished high school were fighting a guerilla war. Psychologically, many of them were not ready for this and suffered from the effects for decades to come.

President Johnson ran as the "Peace Candidate" in 1964, declaring, "We are not about to send American boys 9 or 10,000 miles away from home to do what Asian boys ought to be doing for themselves." Within six months after his inauguration, President Johnson increased the draft from 17,000 per month to 35,000. This, in turn, fed the anti-war movement that was growing in America. It was unfortunate that part of this growing anti-war movement was directed toward the American service members who were serving in Vietnam. The overwhelming majority of us were doing our jobs and honoring the rules of warfare. Soldiers of our platoon obeyed the rights of humanity and the Geneva Convention when those two NVA soldiers from the bunker were captured immediately following one of our own being killed by a sniper.

Retirements of senior NCOs and officers who had previously served in World War II and the Korean War were unavoidable. For over a decade, promotions into the senior enlisted ranks had been very slow, and I was one of the few young ones to work through that problem by studying hard and graduating top of my class in Ranger School. Had it been just a peacetime process of replacing the retiring NCOs from the existing ranks, our brigade and subordinate commands could have made it work.

One of the major problems we had with platoon leaders and commanders was their three-month rotation cycle. That hurt us badly. Within my year tour, I had four platoon leaders. Soldiers had to stay in that squad or platoon for a year, while it appeared that lieutenants and captains were going through a revolving door. A bigger part of this problem was that senior commanders tried to get combat leadership time on junior officer resumes. At about the time junior officers completed the three months in the combat indoctrination program, they were rotated out. The best of them had built professional bonds with their soldiers. Attempts on their part to stay were considered in the way of another junior officer's opportunity to "punch a ticket." It was their ticket against our lives and the command's ability to perform its mission. Fortunately, today's Army avoids that three-month rotation policy.

Because of the wounds and subsequent disabilities, a lot of good NCOs were taken out of action. The Army's solution was to quickly build another NCO Corps through what we called the "Instant-NCO Program." The Army sent junior enlisted soldiers to Fort Benning, Georgia, and put them through additional training. They would graduate as sergeants or staff sergeants. They were then sent into combat as squad leaders, often while still serving their first enlistment term. Many of these "shake and bake" NCOs had never been in a real unit established by the Modified Table of Organization and Equipment. Now they were supervising junior sergeants and specialists who had experience, proper training, and had participated in combat exercises before coming to Vietnam. Who should have been the squad leader? Probably a young sergeant with some experience rather than the instant-NCO with none. That really caused a problem because the "shake and bakes" were making mistakes that soldiers who worked their way up the ranks would have avoided.

You cannot blame the individual; the system was the problem. Even that great soldier Specialist Jesse James was denied proper training. He was sent back to Phan Rang, provided a week of training, and made a junior sergeant. At least his soldiers received a warrior who knew combat patrolling and had a sixth sense for how to keep them alive in a field of fire. Yet, there was so much more to NCO development that he should have been provided before putting on three stripes. The same thing happened with the officers. Draftees who did well on the Armed Services Vocational Aptitude Battery and had a couple years of college were sent to OCS followed by Officer Basic Courses for their specialties. Just months after leaving civilian life behind they were lieutenants in combat.

That said, there were thousands of these early advanced soldiers who did an outstanding job and served their country well. A lot went on to become career NCOs and officers. Once they returned to the States, they were able to get a proper professional education. Unfortunately, many young officers did not pursue their civilian education requirements by earning college degrees through night schools. In all honesty, that was on them. Every base had a night school program in which accredited university degrees could have been earned. They could also have taken the College Level Examination Program test to gain credit for at least a year of college. The test results were accepted by all universities operating extension programs on base. Additional credit could have been earned through recognition of the military courses. Stateside, in Germany and Korea, night courses were designed so they could complete a year's worth of advanced college study within a year.

Thousands of officers mistakenly believed their combat records would carry them through the board review process. Instead, their lack of education became the first determining factor for retention. Many good officers were caught up in the Reduction in Forces (RIF) in 1973 and 1974. They either left the Army entirely or were reduced to NCO ranks in a repeat of the 1950s and '60s, where the NCO ranks were jammed up, restricting high-performing soldiers from advancing. The adage is true: "Those who fail to remember the lessons of history are destined to repeat them."

On the opposite spectrum, those who remember lessons received are destined to benefit from them. This realization proved itself time and again in Vietnam. In our professional development schooling, I was taught things that were valuable in combat. These things went well beyond the use of weapons and equipment. Instructions included the best ways to effectively work with people. One of those lessons was applied to the replacements as they came in. Ranger School had taught us to use a buddy system. Two people were officially assigned as buddies. Even though there may have been a personality conflict, they still had to work together. We did the same thing in our platoon in Vietnam. A new replacement was assigned to an experienced platoon member who had been there about six months. They would become buddies, and the replacement quickly learned how to execute his responsibilities. The buddy system usually works in any type of organization. Putting two people of equal experience together is not the way to assign buddies. Everybody should want to do the best they can and help their buddies. Buddies are the first level of a team; then teams become squads, platoons, and beyond. There was no "I'm doing it just for me, but not for the team." That is a critical mindset in any military operation, especially combat.

Another applied lesson was that all soldiers had to earn their way. For example, before a soldier was allowed to serve as point man, he had to prove to his supervisors that he could do the job. The rest of the squad and platoon were going to walk into the terrain the point man had cleared. We could not risk getting the platoon shot up because the soldier in the front was operating beyond his skills.

Professional soldiers try to do the best they can to accomplish the given mission. In a firefight, there is one objective: to defeat and destroy the enemy. Consciously and subconsciously, leaders are continually conducting battlefield and risk assessments. This is not by the minutes, but by the seconds. A leader also must have trust that all team members are going to do their jobs. Realistic training is critical in preparing a soldier to become an effective combat warrior. There are major differences between the two. When contact is made with the enemy, warriors fight as they were trained to fight. At all soldier levels, there is a lot going on. Each warrior has his own area of responsibility. A squad leader moves people around while communicating with all subordinates, ensures soldiers are not wasting ammunition, is always on the lookout for weak points that the enemy can exploit, works closely with other squad leaders, stays connected with the platoon leader and sergeant, and works with them when in need of supporting fire—all while exchanging gunfire with the enemy.

In the previous case of C Company's battle, we would take that bunker regardless. The squad or platoon leader could not suddenly stop and render first aid if we took casualties. That was the medic's job. If the leaders and team members lose their mission focus, the mission will be doomed, and more casualties will occur. We took that bunker and, in turn, secured the open field because we stayed focused. The medic did better than we could have done for our fallen soldier. If a soldier had any chance of survival, it would have been stripped of him if we could not get a medevac helicopter in. If we had allowed ourselves to be overwhelmed at any point, that bunker and the open field would have remained under enemy control. Leaving a known enemy in our rear was not an option as we proceeded to advance.

One of the best actions of our leaders back at Fort Campbell was mixing training locations. We did not become complacent by always training in the same fields and forests. Whether we were maneuvering in Tennessee, Kentucky, or West Virginia, we had to adapt and learn in different environments. We did not know what was over the next hill other than what our maps identified. Also very valuable was working our way through the canopy coverage in Georgia during Ranger School. Thank God I had the experience and was well-trained in those areas.

Any command that uses the same territory repeatedly to train their soldiers is doing no one a favor. That includes the logistics and other combat service support commands. In the short term, repetition of using the same territory time after time makes it easier for them to work from long-established supply channels and depots to support us. However, it undermined our effectiveness in the long term. Combat does not seek out the easy way; if anything can go wrong, it will.

CHAPTER 7
Return to the States

When my first year in Vietnam was over, I had conflicting feelings about leaving the soldiers I had trained and led who were staying behind until their year was complete, provided they were not medevac'd stateside or killed in combat. We were a team in which each member took care of everyone else. One option for me was to stay another year, and the second option was to return home to my family.

The families are the ones to endure the never-ending uncertainty. When not engaged in combat operations, I knew it and could relax. If we were preparing for combat operations or involved in a firefight, I was too busy to give anything else a thought. Margaret had to carry the entire family load while I was gone, and a single parent raising a child has a tough road. Someone always wanted to take advantage of the situation, and Margaret always had to be on guard. Each night, when she went to sleep, she was thankful that a taxi driver had not shown up or left a Western Union message on our trailer door stating that I had been killed or wounded in combat. It was time for me to return home. Notification came that once this combat tour was complete, my next assignment was to serve as an instructor at the Ranger Department at Fort Benning, Georgia. For all my school training and real-world experience, I was being called upon to share it with others who would most certainly be assigned to Vietnam. The instructor position would allow me to care for my family while helping other soldiers stay alive and achieve future combat success.

The first stop in coming home was a transition center at Fort Lewis, Washington. We were there for a day and a half. We were issued a new dress uniform, two sets of fatigue utility uniforms, underwear, socks, and something we had not seen in a long time: new footwear. Then came a hot shower, the best in a year. The staff worked with us to arrange transportation back to our homes. My connector flights from the Seattle-Tacoma airport to Nashville gave me time to think about a lot of things: what was accomplished, what could have been done better, and what was ahead.

Over the years, I have reflected on how fast I went through Fort Lewis. More time should have been spent on us soldiers, giving us time to decompress and adjust to not being shelled and shot at and now being in a peaceful environment. The Army made a mistake. All experiences in life's continuing development result in personality and behavioral adjustments. Major events and intense environments, such as active combat operations, will have an even greater impact. Many factors go into how an individual adapts and whether the experience has a positive or negative influence.

Personally, I had several advantages over other soldiers going to Vietnam. From my earliest memories, growing up on a farm meant a lot of hard work, and giving of yourself to look out for each other. Later, the Berlin Brigade allowed me to develop as a professional soldier. Our convoys through East Germany taught me the fallacy of communism. Airborne School, Ranger School, and being trained by leaders like Sergeant Franco contributed positively to my development. I had time and a lot of great leaders to help me physically and mentally prepare for the challenges of an aggressive combat tour. Many who either volunteered or were drafted shortly after completing high school did not have the opportunity to develop professionally. Processing through Fort Lewis as quickly as I did was acceptable because I was ready to return to my family and begin my next assignment. That did not mean I liked what I witnessed; as for others, there could have been more opportunity for decompression and a chance to be identified if more transition support was necessary.

That shortfall contributed to the difficulty many Vietnam veterans endured once they returned stateside. For some, it is a lifetime challenge. Decades later, veterans coming back from Iraq and Afghanistan received additional support, evaluations, and counseling. Even then, with the high suicide rates among returning combat veterans, success has proved itself to be elusive. One may only hope that this process will be increasingly perfected.

Margaret and our daughter Melissa were waiting for me when I stepped off the plane in Nashville. Margaret was everything I remembered. When I left, Melissa could only move around on her hands and knees. Now Melissa was walking and talking. She claimed she knew who I was, but I was unsure. With them was my dear mother-in-law, who was always there to help us in every way possible. My doubts were erased when we got in the vehicle. Melissa was in the back seat with her grandmother. Margaret was with me in front. Melissa said, "I want to get up in the front seat with my dad. That's my daddy." I guess she was proud of her daddy. Our family was together again. In June 1967, we arrived at Fort Benning. The move only required us to load up the car. The Army reimbursed us for the trucking company that hauled our mobile home to Georgia.

Today, the Ranger Department is the Ranger Training Command. My instructor position involved teaching hand-to-hand combat, basic patrolling techniques, land navigation, and physical fitness, including enduring road marches. About half of the instructors had served in Vietnam. The other half referred to themselves as "professional instructors." There is a saying, "Those who can, do. Those who cannot, teach." Claiming a title like "professional instructor" has its limits. When possible, there should be a requirement for combat before someone is assigned to train others for war. In the early days of what will become an extended conflict, requiring instructors to have combat experience can be a challenge. Once the conflict has been going on for a couple of years, there is no reason to have peacetime homesteaders trying to tell others how to fight. It is for the instructor's own good as well. If someone turns into a professional instructor, including drill sergeants and Military Occupational Specialty (MOS) trainers, they will have problems when they return to a combatant command. The skills they should have developed while moving up the ranks will not be there.

It is different applying skills in a command that has multiple things going on simultaneously, while taking care of soldiers and equipment. A maintenance instructor can teach how to fix a vehicle and order replacement parts. Should that professional instructor be assigned to a company or battalion command where the equipment is broken, and replacement parts will not arrive for months, he or she probably will not know how to find the parts and get the equipment repaired. They only know the books and their lesson plans. This is especially true in combat. Meanwhile, a good, experienced motor sergeant knows how to maneuver through these problems because he has been doing it for years.

If they ever get back to combat deployable units and these professional instructors cannot quickly adapt, they will be totally lost. They are going to drag their section down with them. Subordinate soldiers will immediately pick up on this. Respect and trust are hard to rebuild once they have been lost. This is even harder when the person expected to lead is lost. The best option is to be in a training command for a couple of years, then go on to bigger and better things. Over time, I have witnessed those who could adapt and go on to be successful. Those who could not would fail.

The initial phase of the Ranger Course lasted three weeks, and there were no weekends off. Days started at five o'clock in the morning and finished fourteen or fifteen hours later if we were lucky. The instructors and support staff received about a week's break between class rotations. Sometimes, we were able to enjoy a five-day weekend. We did not need to teach specific courses on psychological endurance. The pace we kept did that for us. Those who could handle the pressure and met all the requirements graduated from the course. Those who couldn't were returned to their units.

Obstacle courses had a different focus because everyone had to complete them. If someone could not keep pace, the rest of the command did not leave the soldier behind. The students benefited in two ways. The stronger looked out for the weaker. The weaker had to push harder not to drag the team down. The instructors were watching to see who needed further evaluation. This not only applied to the weaker but to those more interested in making themselves look good at the expense of others. In combat, both types are problems and were dealt with in our training.

During the first two weeks, the students lived in World War II platoon bay barracks. In the third week, they went to a small camp down the road and lived under their ponchos and conducted patrolling. This built upon their already-existing skills. After our basic phase of Ranger training, the students moved on to the mountaineering and Florida phases. They trained in the big swamps, walking around in water up to their necks while on patrol. The Florida phase also included the escape, evasion, and interrogation course.

It was important for all of the instructors in each progressive phase to toughen the participants up even more or weed them out. We did not want them to fail, but it was even more important that none died because we were too soft on them. Whether the students were officers, NCOs, or junior enlisted, they all had to achieve the same standards to pass.

The students we were training were on their way to Vietnam, if not within a year, within the one after that. The poorer the training they got, the higher the risk of their deaths and the deaths of those around them. Especially with the combat veteran instructors, we were determined to influence others so they could go there, perform their missions, and bring themselves and their fellow soldiers back home safely. That is what training was all about—being able to fight a war, come home safely, and afford the same opportunity as their battle buddies. We also shared many of our experiences of being able to navigate very thick terrain. The students really appreciated those of us training from experience.

Understandably, the summer courses created the most heat-related problems. Most of these were dehydration from not drinking enough water. We also had a lot of foreign students who seemed to have more problems with physical fitness than American students. At Ranger School, we were allowed to use our imagination and combat experience to make training more realistic. Using that flexibility, we built our own VC village out of bamboo. To add realism, we added live chickens. My buddy instructor took it further, bringing live hogs into our mock village. Having grown up on a hog farm, he knew how to care for them. Whatever his first name was, he was soon known as Sergeant Hog. Those critters added to the realism of training as we did encounter hogs and other animals when patrolling through the Vietnamese villages.

Sergeant Hog collected all the garbage from the dining facilities and had a pretty good operation by finding a way to continue his family's passion for Fort Benning. The dining facilities avoided sending their garbage to a rat-infested landfill. When the hogs became full-grown, they were sold, and young ones came in as replacements. Also, we never had to be concerned with snakes in that training village. Those venom-filled reptiles were nothing more than morsels to those hogs. As with my family's hogs in North Carolina, these hogs were intentionally grown fat. A snake could never bite deep enough to get its venom past all that lard. Sergeant Hog's enterprise was a total win all the way around. Nowadays, there are probably a dozen regulations and laws that would prohibit such an operation. This did not hurt Hog's career. He was accepted into OCS and eventually retired as a major.

We had a lot of fun training Ranger students, putting on a lot of hand-to-hand combat demonstrations, rappelling demonstrations out at Victory Pond, and practicing coming down a rope into the water during the slide for life. Hog and I were teamed as hand-to-hand combat demonstrators. In front of the audience, we would perform different full-contact maneuvers on each other. All our acts were staged, including pretending to get mad at each other. I would kick Hog, and he would holler like I had whomped him too hard. He would get up, frown, back away a little bit, put his hand on his hips, throw me and work me over well. I would get up, twist my neck to set my head, grit my teeth, and spit. We would rotate back and forth about who was winning—same as all-star wrestling.

The role of our moderator was to keep the audience from getting involved and trying to separate us because they thought Hog and I were really getting into a personal fight. We were just two country boys having fun. At the end of it, I would throw Hog and kick him until he turned over on his back. Then I would jump down on his chest and take my thumbs like I was gouging out both his eyeballs. Hog would play along, distracting the audience while I slid two grapes into my hands. At the right moment, with Hog covering his eyes, I threw those grapes into the audience. The troops went nuts.

One day, we had some British generals in the audience who thought we were not in contact with each other. After the demonstration, our colonel said to them, "Let me tell you something, gentlemen, there's nothing phony about this. If you don't believe it, I will show you something. Come here, Sergeant Gates." I walked over to him and stood at attention. The colonel pulled up my T-shirt. There were blow marks on both sides of my body. One of the bruises was bleeding a little bit. One British general said, "Ah, you Americans are crazy."

Soon, I was going to be working with the British on their soil and have further opportunity to convince them that Americans are, in fact, crazy. I was selected to attend the three-month British Army Tactics Course in South Wales. The British soldiers asked many questions about what was going on in Vietnam and how we were training. We established good relationships. Even the instructors appreciated the opportunity to increase their knowledge. They did their best to maximize the utility of my experience. They would start with a simple scenario and then build as the conversation progressed. One example concerned dealing with prisoners of war. We would start with the capture of two prisoners while on a mission. We would address how to properly secure the prisoners so they could not escape or compromise our mission. In further building their scenarios, the Brits would add two wounded or sick soldiers to our mission.

The discussions progressed with increased severity of the wounds or illness and the urgency of getting medical treatment. Caring for the downed soldiers could result in making a litter (stretcher) out of a poncho wrapped around two poles or even carrying them on our backs. This could result in a two- or four-man carry. Not only were the wounded removed from being combat effectiveness, so were the transporters. Eventually, the scenarios developed into the solution, which was to find a site to medevac the wounded out, along with the prisoners.

As the British cadre built upon these scenarios, I was usually held in reserve until the other course members explained their proposed solutions. The cadre did not provide answers; they had the students work their way through the various proposals and discuss solutions with each other. This method enhanced team communication skills. When they came to their conclusions, I was called upon to explain how we did it in Vietnam. But just because we did it in Vietnam did not make it perfect. Sometimes, the British discussion groups came up with excellent solutions when the students developed possible ways for field warriors to do it better. Some of their ideas became stored in my memory bank and were available for use on my next tour of duty in Vietnam.

For about a week, we did map training for combat cities. To us, it was MOUT. The British Army maintained a town evacuated during World War II to practice that. Everything was still there. Because of their experience in Northern Ireland, this is where British training far exceeded American. Concerning MOUT, our instructors trained from what they had studied. The Brits were training from what they had lived.

The British Ranger Course was tough. Their abbreviated physical fitness test included push-ups, sit-ups, and other events, followed by a three-mile forced march. Unlike the American hard steady walk, what the Brits call forced march is walking as fast as possible to get to the top of the high ground, then running down the other side. The full physical fitness test consisted of a ten-mile forced march followed by an obstacle course. The close-out at the end of the obstacle course was shooting five rounds from a rifle carried through the entire test. The passing requirement was to make three hits. No matter how out of breath we were, with the proper use of basic shooting principles, we hit the target all five times.

Another of our PT events was the combined construction of a defense encampment followed by an escape and evasion course. We dug foxholes, trenches, and connecting trenches for three days. At night, we would conduct patrols. In much more forgiving terrain, and among a much more friendly population, this was like recondo missions in Vietnam. After the third night of patrol, we rendezvoused with our trucks to return to the base camp. Everyone was exhausted and looked forward to a couple of hours rest before we started digging again.

Soon, we were advised that the camp had been overrun. The escape and evasion portion of the exercise was on. Our mission was to go cross country to a safe haven about ten miles away. An opposing forces (OPFOR) regiment had been assigned to capture and deliver us to an interrogation unit. We split into small teams. Having done similar training in America, I told my five team members, "This is how this works. When people cross that line, they're going to capture you, so what we're going to do is get beyond that line before they form."

Looking at a map it was easy to figure out where the OPFOR was going to be establishing itself. For the next five hours, we ran at a sustainable pace. If we slowed down, it was only to a forced march. We ended up at a little village that night and took refuge in a smokehouse. No longer being used to smoke foods, it had become a little barn for storing canned foods. We could hear the OPFOR outside, walking around looking for us. Perhaps because it was private property, the smokehouse was never searched. Eventually, the OPFOR moved on.

The next morning, the lady of the manor came out. She opened the door, threw her arms up, and said, "Oh, my God, what are you boys doing in my place?" We explained to her what was going on. She asked, "Can I make you a cup of tea?" We accepted. She brought us some nice hot tea and food. She was old enough to have remembered World War II and the importance the European Resistance and Underground played in getting escaping British and American service members back to friendly lines. She got into the act. We stayed in her sanctuary all day. As soon as it got dark, we thanked her and moved on. Having had time to rest and be fed, we were recharged.

We made good distance over the second night and stayed away from the patrols. As night ended, we found a cave where sheep stayed to get out of the weather. Sheep are nonaggressive animals. If not for the fact that they drop their waste out of the back end as fast as they eat through the front, they could almost be domesticated. While the sheep were coming and going to graze in the field during the day, we took up residency at the back of the cave. Either the OPFOR did not know about the cave, had no desire to step through the sheep droppings to come in, or thought no one would be crazy enough to hide out with sheep. When we left, we must have been ripe.

That night we resumed our trip and linked up with friendly forces. The British were upset because they had military intelligence people assigned to interrogate the ones they captured. The five of us were the only students not to get captured. We returned to our base where the colonel and Regimental Sergeant Major (RSM) were waiting. The RSM said to me, "We really wanted to capture you because we wanted to practice interrogating someone other than a Brit." Then the colonel asked, "Would you volunteer to go and let the military intelligence people interrogate you anyway?" I agreed to his request. The RSM spoke up. "No, Colonel, the damn sergeant cannot do that. He won the right to not be interrogated." The RSM then told me I had saved him from paying out a bonus. He had posted a bounty of twenty pounds, payable to the person who captured the American. Back then, twenty pounds was a significant bonus to someone living on a soldier's pay, be they British or American.

I was not a marathon runner, but I was in top physical shape during my time in the British Ranger course. Running ten miles and carrying all the equipment was not a problem. One time, while we were doing a forced march, two or three classmates had fallen back. I called back to them, "You cannot keep up with a Yankee." While I was saying this to the stragglers, the colonel came up and asked, "What do you think about this march style, Sergeant Gates?" I said, "Colonel, with all due respect, I think you need to speed this thing up a little bit. It's going entirely too slow." The colonel replied, "You're the craziest man I've ever seen."

He was the second senior British officer to call me crazy. Both may have been right. Part of it was American bravado, the mindset to appear as a slightly off-balanced maverick. Our nation was built on people coming from all over the world to achieve success by breaking with traditions. If Andrew Jackson had fought like the British at New Orleans, the American defenders would have been slaughtered. Instead, he put riflemen behind cotton bales and had pirates manning cannons to shoot down the marching British columns. Fortunately, America and Britain settled their differences in the 1800s. Since then, we have always been the best of allies. My experiences with them in Berlin and at Ranger School were enjoyable. They are hard workers, make fine soldiers, and have an interesting sense of humor. They can always laugh without losing a bit of their professionalism.

As the end of the third month approached, it was almost time to return to Fort Benning. To the British standard, they could not send the American back without a unique experience that I never had before and would never have again. They had to make available a special memory, being teamed up with a parachute regiment for five balloon jumps. The Brits were right. Those were the only times I jumped from a balloon. While my experience there was great, it was even better to be back with my family. Returning to my instructor duties at Fort Benning was even better with the promotion to sergeant first class. The promotion list had been out for about a year. Finally, the approval orders for the fifth stripe came through.

Family life went very well despite the long hours instructing at the Ranger Course. Miss Margaret was working as a cashier at one of the first K-Marts built in Columbus. The pay coming in from the Army was adequate for us to live comfortably as a family, but she wanted to contribute and help support the family. The extra money helped. Equally important, Margaret had come from a hard-working family herself. She continued her family tradition and felt good about doing it. Everything about Miss Margaret and everything she did just made me love her more.

Despite our workloads, we were still able to get involved in the community and enjoy Georgia. One off-duty hobby was volunteering as a Cub Scout leader. It was a lot of fun working with the Webelos (4th and 5th graders). Another hobby that took me back to my own youth was hunting. The difference was I had moved up from squirrels and rabbits to hunting deer. One thing never changed. Everything shot ended up on the dining table; if not ours, then someone else's. Some parental lessons last a lifetime.

CHAPTER 8
Spirit of the Bayonet

It had been about a year and a half since I had finished my first combat tour when, near the end of November 1968, notification for my second Vietnam tour arrived. I said farewell to the Ranger School in December over the holidays and prepared myself and my family for a January departure. This was with us, knowing that I would be leaving Margaret and our young daughter behind for another year. In that short time, since I had departed, the entire direction of the Vietnam War had changed. When I was last there in 1966, our job was to kill, capture, and destroy the enemy. During the waning months of the Johnson Administration, the directions coming down from Washington were for pacification and country-building. The focus had shifted away from offensive operations to an emphasis on units defending their bases.

Departing out of Oakland, I was transported again by commercial airlines to Saigon. Into Vietnam our flight landed at Military Assistance Command–Vietnam. From there, we received transportation to our unit, and I was back in the central highlands. The company had a mission for long-range reconnaissance along the Cambodian and Laotian borders. We were attached to the 4th Infantry Division (ID), based at Pleiku. The base had semi-permanent wooden buildings where soldiers could spend time decompressing. Positioned further into hostile territory the brigades were in tent cities. These temporary bases maintained perimeter defenses with firing positions that were either built into the ground or in fortified bunkers.

To the west, a brigade of the 4th ID was operating out of tents near Con Thien City. While the brigade operated primarily in the defensive mode, the Ranger's mission was to conduct deep penetration patrols to detect enemy activity that could endanger American and South Vietnamese forces. Unfortunately, senior commands viewed our missions as secondary to the pacification and country-building emphasis.

Whether by day or night, the missions were normally done by insertion. We intentionally mixed up the times so enemy observers would not know when to look for our activities. The best time was during the darkest hours, which gave us the most concealment, and when NVA soldiers were more likely to be asleep. The ideal condition for insertion was during bad weather. On some of these operations, we did have the option to call in long-range indirect air support to interdict enemy movements. The patrols did this quite often and effectively.

Sometimes our teams would be inserted a couple of thousand meters away from their destination and maneuver to the actual operational area. Each team consisted of a leader, the radio operator, and two other members. They would normally stay out for about five days, living out of their rucksacks and using coded communications to stay in contact with the base camp. Unless there was some type of emergency, we operated totally in code. Our job was to provide surveillance and reconnaissance in certain areas along the border in support of the 4th ID mission.

Most of K Company were good soldiers. Some were Ranger-qualified, and all had completed airborne training. Combat experience was a mix, ranging from those who had completed at least one combat tour to soldiers experiencing their first. Whether drafted or voluntarily enlisted into the Army, most of the soldiers in K Company had volunteered for their positions.

The jump-off point for our missions was a small mountain village named Pleimoron. There was a joint camp consisting of an artillery battery and a Vietnamese Special Forces camp, with four American advisors. A side benefit to jumping off from that camp was that my platoon and the artillery troops got to know and appreciate each other. When artillery engagement was necessary, they were real people with known faces on both the shot and splash ends of those fire missions. The Ranger patrols and artillery crewmen were teammates.

Our standing orders were to avoid contact and move quickly and quietly over long distances. Lugging heavy machine guns and ammo would have restricted our mobility, which made it impossible for the four-member teams to carry M60 machine guns. If there were contact, we would immediately get that team out of the operational area because its mission had been compromised. Generally, the enemy never knew the teams were in the vicinity.

For each operation, my soldiers would be brought in from Pleiku. We would go through the planning phase and ensure they were fully equipped. From there, they would be inserted into the Chu Pong Mountains near the Cambodian border. Once their patrol was complete, they would be transported back to Pleiku for rest and recuperation. During one operation, intelligence reports informed us that the NVA was possibly planning to pass through Chu Pong, traverse the mountains, and launch an attack on the 4th ID's camp. Our insertion teams were ready and did detect a large NVA element working its way east; however, that NVA unit was decimated by Ranger-coordinated artillery.

On another day, a team reported that they had been detected and were conducting evasive maneuvers. The way they got out would be discounted as pure fiction if seen in a Hollywood movie, except it was real. Moving down onto the low ground, they took refuge in a stream and stayed under foliage for a couple of days, sometimes submerged and having to breathe through a reed. Through radio communications, when they could send them, we knew their exact location, but rain prevented us from getting to them.

When the rain let up, a light observation helicopter (LOH) crew took on the extraction mission. That pilot went right into their location and hovered above the water. Those guys grabbed hold, crawled up, and straddled the helicopter skids without any straps. There was no room inside the LOH as there was only room for the pilot and copilot. It took a great flight crew to do that, and one with a lot of intestinal fortitude. Those soldiers were still riding the skids when the LOH came into the camp at Pleimoron.

Back at base camp in Pleiku, K Company's first sergeant and I were walking toward the 4th ID headquarters for a mission brief from the intelligence and operations sections. We received a radio call that one of our deployed teams was in contact with the enemy, and they had been compromised. We were also informed the team leader was seriously wounded. The mission brief would have to wait, and we detoured straight to the aviation command.

We told the operations officer, a major, of the situation. The unit executive officer was also present and said, "Well, we will just have to wait and see how it plays out." The operations officer responded to his senior, "Goddammit, we are going to go out there, and we will get all the damn helicopters going out there. Sergeant Gates, if you want to go, we will do it." There was only one answer; they were both my troops and the first sergeant's. Not only did the major order the mission, but he also flew it. He and his copilot brought their helicopter in as close as they could to the compromised team, which was three hundred meters away. From continuing radio communications, we learned the team sergeant was dead, and Specialist Gomez had taken charge. One other soldier was wounded, and there were only two people left fully combat effective. Gomez was calling in indirect fire on the enemy locations. When the helicopter landed, I said, "Damn, we do not even have a weapon." The requirement at base camp was to have all weapons locked in the arms room, so the major gave me his pistol. The major was in communication with other pilots while en route, and cavalry support was on the way, but we did not have time to lose. After the first sergeant and I disembarked, the major took the helicopter back to the air because only a stranded pigeon stays on the ground. A soaring hawk does its best fighting in the air. The major maneuvered the helicopter, allowing one door gunner to support the movement of the first sergeant and me while the other gunner supported Gomez.

The first sergeant and I linked up with the team. Working with the uninjured soldiers, we brought the deceased team leader, the wounded soldier, and the equipment out. The first sergeant and I were now armed with the M16s of the fallen. We commenced working our way back to our landing zone in what was no easy task. The NVA were staying in contact knowing the helicopter would have to come back down to pick us up. They were looking to attack us once the helicopter came in for the rescue. Suddenly, a cavalry squadron from the 4th ID came in with attack helicopters and took over what instantly became a one-sided fight. Coming back to the ground, the flight crew got all of us on board, lifted off, and headed back to base camp.

What the pilots, the first sergeant, and I did may have appeared risky to some, but our actions were typical of a warrior. Our soldiers were under attack, and if they were not immediately extracted, it is likely they all would have died. The major piloting the helicopter was a hero to do something like that. He violated a whole bunch of regulations and was insubordinate to his senior. The fact of the matter was that he never forgot his roots. If he could do anything possible to save an American soldier's life, even if it meant doing damage to an expensive helicopter or himself, as a warrior, he would do it. Meanwhile, his crew never hesitated to be right beside him. I doubt any of them would have wanted to go on a combat mission with the executive officer.

When we flew out, the radio and all the essential equipment were with us. The only thing that we could not carry back to the helicopter was a rucksack filled with rations and a poncho. Some NVA probably got meals for that night out of it, if there was anything left of the rucksack and them after the attack helicopters were done with that area. During the debrief, we were chastised by an officer for leaving that rucksack behind. I was still fired up. My reply was, "If you had been there with your mentality, you probably would have left the radio and wounded behind, too," I gave him the coordinates and said we would take him back to that same place so he could get the rucksack. I then told him, "I will pay for the rucksack if that is the only thing on your mind. I will pay for the damn rucksack." He did not accept my offer to be flown out to the battle site. It was easier for rear-echelon staffers in base camp to complain than to get out and do something, especially when it meant going into harm's way. The sad part is, when his deployment was over, that person who was more worried about a rucksack than a human life would be strutting around with his combat patch and Bronze Star. Everyone has a job to do, but those in the rear echelon should have also understood what combatants endure. If they do not understand that, they have no business giving orders to the soldiers who are outside the perimeter fighting, killing, bleeding, and dying.

When you beat the enemy, you feel good about it. When he beats you, you live with it. The lives the first sergeant, the helicopter crew, and I saved were overshadowed by the soldier who was lost. The first sergeant and I ensured the wounded soldier received proper medical treatment and that the fallen soldier's remains were delivered to mortuary affairs. A letter to the family needed to be written, but that could wait a day as it would take a week for the family to receive our words by mail. Because Vietnam's time zone is many hours ahead of the United States, the family would learn of their son's death before their day was over.

By the time things settled down, it was evening. The first sergeant and I went to the NCO club to unwind over a cold beer. We were in the same jump uniforms that we wore when pulling our soldiers out of the firefight. We still had blood from the deceased and wounded soldiers mixed in with soil stains picked up while rescuing the team. At the NCO club, the door monitor would not let us in because our uniforms were not "pressed and starched." I asked him if he "ever had a Ranger ass-whipping." He went and got the division command sergeant major, who came running over and told me to come to parade rest. My first sergeant responded, "Sergeant Major, have you ever had your goddam ass whipped?" The sergeant major got two words out, "You ain't . . ." My first sergeant cut him off. "No, I'm threatening you. I don't want to hear this shit. Sergeant Gates and I are coming into this club, or we'll shut the whole damn thing down." A pin-dropping could have been heard in that club. Suddenly the entire place erupted in applause and cheering. With the entire club backing us, we got our beer. It should not have had to come to that.

Basically, there were two different armies: the haves and the have-nots. The haves were living the plush life in hard barracks and hotels while warriors in the field were in poncho tents. The most dangerous patrols they went on were to the post-exchange, the dining facility, or a local entertainment facility in a nearby city. It was an unavoidable sin that soldiers fighting the fight were getting the same combat pay as those who never heard the sound of a hostile bullet their entire time in-country.

Not all the attacks on us were from the NVA or VC. We had a team from our platoon in the mountainous area, at a place we named Sugar Loaf Mountain. From the highest elevation, there was a thousand-meter ridge line that gradually descended into a valley. The entire area was a hot spot of activity. An operating team radioed in that they had been attacked by a tiger, which almost killed one of our soldiers. In their concealed position, that tiger had smelled them out and came in for a meal. The tiger took off after being shot and no one knew if it died. We extracted the team because their position was compromised, and the wounded soldier needed immediate medical attention. Because he was not wounded by enemy activity, the soldier was not eligible for a Purple Heart medal. Unlike the previous soldier who put himself in for a Purple Heart over a leech, this soldier was nearly killed by that tiger and deserved some form of recognition, which he never received.

For a period of time, our operating sector included the Ia Drang Valley. Although the NVA received a heavy bomb pounding after its fight with Colonel Moore's battalion, they had reoccupied the area with a regiment of troops. The Ho Chi Min Trail ran through that valley. There is a misconception that the trail was a single passageway, but it was actually a network of interconnecting roads built by the North Vietnamese. They went to great endeavors to build it. In some places, they had laid wood on the ground to allow their vehicles to pass through the swampy area. Because of the foliage, aerial surveillance was very limited.

One of our missions was to maintain a surveillance network that ran through our mission sector. On occasion, we did find the need to call in artillery. When my platoon received a new mission, we were replaced by another platoon. That area was too mission-critical to deny keeping it under surveillance. A river ran through that valley and three old French forts were built within it. Even though the buildings were made of bricks, everything was in ruins. We conducted a reconnaissance of one compound to determine if a co-located field we observed from the air could be used as a landing zone. Somehow, by the grace of God, one of our soldiers saw a mine. Upon closer inspection, they realized the entire clearing was totally mined. The NVA had prepared well for the possibility of a large force coming in by helicopter. Fortunately, we did not land in the trap.

A platoon-level operation we performed resulted from a briefing that a trail network was suspected of being where the North Vietnamese were coordinating logistic movements with the VC element in the region. Identified by helicopter observation and confirmed by our insertion teams, the trail network did exist. At the briefing, I recommended that instead of using K Company solely for its long-range four-member insertion team reconnaissance missions, we make it a larger operation. The recommendation was approved as a six-team mission.

We worked our way from base camp to the trail site. Once there, we set ourselves up to conduct an L-type ambush. At about one o'clock in the morning, approximately fifteen people came walking up the trail. We hit them hard and when the ambush was over, not one of our soldiers sustained the slightest injury. Surprisingly, there was one survivor from the other side. We brought him back as a prisoner of war. It turned out he was a high-ranking officer and the North Vietnamese logistics representative for that entire area of operations. I did not attend any of his multiple interrogations but was advised that the intelligence community found him to be a major source of accurate information concerning NVA operations in the region. That operation validated the flexibility and adaptability of Rangers in combat.

An important thing about K Company was that we conducted continuous operations. Along the entire border in that area of operations, our four-man patrols would normally be inserted for five days, unless we had to get out in a hurry. The patrols would then return to base, regroup, rest, re-equip, and get a little training before going out on the next operation. We were very fortunate that we did not have a lot of people wounded or killed. We operated twenty-four hours a day, seven days a week, and had people out all the time, even during Christmas. It was very taxing. The rotation of individual replacements allowed freshly assigned troops to be paired with experienced troops.

Christmas was not pleasurable for us. It was not because we were thousands of miles from our families; it was because of what we had to witness going on in front of our insertion teams, and that we were forbidden to do anything about it. Politicians in Washington and Hanoi had agreed to twenty-four-hour Christmas truces which each year gave the NVA opportunities to move huge contingencies of people and equipment without fear of artillery or air strikes. They were able to infiltrate areas they otherwise would have been shot up for trying to do. Going to war and having a cease-fire because of Christmas made no sense at all. When the cease-fires were over, the enemy was well-stocked and well-protected. The Tet Offensive of 1968 was the ultimate example. Tet began in January 1968. During the cease-fire immediately prior, General Giap prepositioned all his troops and all the ordnance. Our covert teams covering the Ho Chi Minh Trail road network witnessed all the movements. The only thing slowing down the NVA resupply was traffic jams. We had all kinds of targets of opportunity. Instead of calling in air strikes and artillery, the best the teams could do was report their observations.

I've talked to a lot of my friends who have returned to Vietnam since the war, and their perception is that the people in South Vietnam are better off today than they were then. The politically unopposed Communist Party of Vietnam controls the government, but the rice farmers have their little farms back and are producing rice. The people are living in a secure environment; they have decent medical care. Electrical power has increased dramatically compared to then. They did not have these things under a supposed democracy that was hijacked by corrupt officials.

Over the years, I had plenty of time to review what happened in Vietnam: what went right, what went wrong, what were important lessons to hold on to, and what actions needed to be done differently. Lessons included understanding tactics to be employed at differing levels. Seasoned warriors understand the things that will work and will not. A well-thought-out operational order must be emplaced before a mission starts. It does not have to be a perfect plan. Another Patton philosophy was "go with a good plan now rather than a great plan later." Waiting for the perfect plan will cost operational delay and allow the enemy to develop their own actions. While not being perfect, a good plan will allow for the subsequent issue of fragmentary (frag) orders in a fluid battlefield. Those frag orders are critical and must be dispatched in a timely manner.

Both operational orders and frag orders must be based on as much solid information as possible. Once those orders are issued, the information within them must be thoroughly disseminated. Soldiers must be allowed to go into a fight knowing as much about their mission as possible. Specialist Gomez knew everything he needed to know before his mission started. When his team leader went down, Gomez was able to keep up the gunfight, call in artillery, care for the wounded, and reach out for backup, all at the same time. Those are things learned in practical application in a real environment when good leaders are in charge.

We must have rules of engagement in any type of war. Mandates on what constitutes a target and when to engage must be published and thoroughly distributed. Leaders must continually ensure that soldiers not only understand those rules of engagement but are properly supervised to enforce compliance. Rules of engagement must be realistic and not impose restrictions on US forces that favor the enemy. Requirements cast upon us obviously came from people who had never operated outside the base perimeter. First, was "no fire zones." There were sectors where we could not shoot our weapon even if we were taking fire from adversaries we could clearly see. Calling for indirect fire in those zones was out of the question. Those rules were implemented to protect the civilian population and prevent collateral damage. The problem was that the enemy quickly figured out which places were no-fire zones and used them as sanctuaries. If you are in a combat zone, every place should be a fire zone. Training and discipline were the proper solutions, not placing restrictions on our ability to survive. There was no answer to the question the soldiers were asking: "Why am I getting shot at, but I cannot shoot back?"

Wars are not successfully fought and won in a military conflict solely based on leadership and training. Logistics are always critical. It is difficult to continuously request and receive needed items at the command level and then ensure those supplies are distributed to the individual units. Leaders should do everything they can to take care of their subordinates, get them the resources they need, and ensure they know how to use them. All leaders and soldiers must understand the capabilities and limitations of their issued equipment. This builds mission readiness, effectiveness in battle, soldiers confidence in themselves, trust in leadership, and command morale.

The squad leader and platoon leader need to understand enemy tactics, their overall mission, and how they go about carrying out these tasks. As for the Rangers and the Special Forces, those are the things that came out of Vietnam with us. It took us a while to get those lessons learned and put in place, but the knowledge we took out of Vietnam we immediately applied in our Ranger courses and training back in the States. We updated our training over the years as more warriors came out of combat.

Unfortunately, many commands failed to recognize that all knowledge is perishable if left only to memory. It is a fact of life that individuals forget the finer points when not exercising what they learned on a continual basis. This was a big problem among the leaders. Rotations meant a year's worth of experience getting on a plane, and departing. Returning to Vietnam an average of eighteen months later, they had to go back to the same learning process. Because of such massive use of the draft and single-term enlistments in Vietnam, all the experience of those soldiers left the military completely. The ability to feel the situation is also a perishable skill. Not all soldiers are like Specialist James whose senses allowed him to detect something going wrong. Even as quickly as James was able to adapt to the environment and recognize hidden dangers, it still took him time. Most soldiers needed about three months to learn how to feel the situation, and some never got good at it.

A great lesson we learned in Vietnam was that there will always be individual replacements, but that is not the way to fight a war. Units should be rotated if they are going to be there for a lengthy period. We made the individual replacement method work, but it was not optimal. A unit should go into combat fully ramped up, trained, and knowing they will go in and come out together. The introduction of team training should not occur while in the line of fire. Strong and weak performers should be identified before deployment. It is a lot easier to do remedial training or exercise accountability before entering a combat zone. It is also less lethal.

Within true warriors is the "spirit of the bayonet." They develop an inner feeling to fight hard, never allow fear to take control of them, and kill when necessary. Some people never get that spirit. A good leader will know who has that spirit. This does not mean going out and arbitrarily killing somebody, it means possessing the fortitude to lock a bayonet onto a rifle barrel and stick it into the chest of an enemy who is trying to kill fellow soldiers or noncombatants. A good leader will know who can be trusted to do the harder right over the easier wrong one hundred percent of the time.

There's nothing more honorable than leading soldiers into front-line combat. This cannot happen in a rear location. There, above anywhere else, leaders learn to respect, and not coddle, their subordinates. It is a rewarding feeling to have confidence in good soldiers, knowing they will do the right thing and fight when needed. What we did have in Vietnam were outstanding service members of all branches going out into the field. The most important thing I learned is that Americans if properly trained, supplied, and led, are the best warriors in the world. They can adjust to any situation. The ingenuity that the average American has is incredible in tough situations where they must invent or create small things to survive. Good warriors blessed with good leaders are unstoppable. That was probably ingrained in me more than anything else.

CHAPTER 9
The Harder Right

Christmas 1969 in Vietnam came with a notification that for the next three years, I would be going to Germany on an unaccompanied tour, meaning my family could not come with me. Going straight from Vietnam to Germany was acceptable, but having to do it without Margaret and Melissa did not make sense. For the previous twelve months, the ladies had done a remarkable job taking care of themselves while dealing with the uncertainty of my safe return. Now it was time to fulfill my responsibilities as husband and father. We had a lot to catch up on.

Returning to the United States came with an automatic thirty days of time off. The division staff in Vietnam provided no help in getting the unaccompanied tour issue resolved. After reuniting with my family at Fort Benning, I immediately drove to Washington, D.C., growing more irritated all the way. At that time, all Army personnel actions were handled inside the Pentagon. This trip came with two learning experiences that never faded. The first lesson was that the building was hard to navigate. Any directions received were either incomplete or hard to follow. Most people got mad, left, and did not take care of the problem they had gone there to address. The second lesson concerned the arrogance of staff members who worked in the Pentagon. Among them were a vast number possessing no combat experience. They lacked any concern for the ones who did and were using their connections within the Army Personnel Office to homestead in the D.C. metropolitan area.

Finally, I located the office where my problem could be addressed. Sitting behind a desk was a master sergeant wearing a Class A dress uniform, revealing a rear-echelon desk jockey. This senior sergeant's uniform lacked a combat patch, and his ribbon rack only displayed the standard "lived and breathed while others did the fighting" awards. Explaining my situation to him, I stated, "I really do not care if the Army needs me to go to Germany; all I want to do is take my family with me." This master sergeant, who never heard the sound of a hostile bullet in his life, responded, "That is the way in the Army. If the Army wants you to be assigned to Germany, then by God, that is where you're going. If you are ordered to go there unaccompanied, then that is what is going to happen." I was in no mood for his arrogance and incompetence. I blew up and was coming around to his side of the desk when a major stepped into the room and asked, "What's wrong, Sergeant?"

My reply was, "Sir, in all due respect, I got to get the Hell out of here or I am going to do something that I have no business doing and will bring discredit on me, the Army, and the NCO Corps." Following up, the major asked, "What is it?" I explained it to him. "If the Army needs me in Europe, I will go. If it needs me wherever—in China, in Japan—I do not care—Okinawa—I do not care. If they need me to go back to Vietnam, I will go. I am a professional soldier. All I am saying is if I am going to Germany, I want my family to accompany me. In the last three years, I have been away from them for nearly two."

The major asked me into his office. He quickly looked at my record and said, "You are absolutely right. This is the end of your second tour. You have your choice of any camp, post, or station in the United States. I am going to give you some orders that will pay for your travel to come up here from Fort Benning. When you get back to Fort Benning, all you need to do is go to the personnel center there. They will have orders on your wife and your child to accompany you to Germany."

Why could that master sergeant not tell me that? Because he had not seen combat. In contrast, the major had been to Vietnam. He had served in the 1st Cavalry Division. He knew the impact of combat on families and the importance of reuniting warriors with their families. There are many ways to decompress after an aggressive combat tour. Being an important part of a growing family is one of them. Time spent with Margaret and Melissa was going to benefit me as much as if not more than them. The major proved to be better than his word. By the time I returned to Fort Benning, the personnel center had already called Margaret and told her she and our daughter had orders. The major had published special orders for my trip, allowing for all travel to be paid for and the days on the road to not be extracted from my leave balance.

After selling the mobile home, we went to Charleston, South Carolina to catch a plane flight to Germany. Upon arriving at the Frankfurt Air Force Base, we were escorted to the train station for a ride to Bamberg. Our sponsor met us when we got off the train. He took us to our temporary quarters where we stayed until our permanent quarters were available. The apartment was even stocked with a couple of days' worth of food. As nice as the temporary quarters were, the permanent ones allowed us to spread out even more.

Germany had changed for the better in the ten years since my days in the Berlin Brigade. The pains of war and the tyranny of the Nazis still haunted the population, and people who had supported Hitler still had to live with their shame. They were getting older, and an entire generation born after the conclusion of World War II was now in their early adulthood. German determination and attention to the smallest details had rebuilt a strong economy. Germany had become America's and the United Kingdom's strongest ally on the European Continent. A critical change had also occurred inside the Warsaw Pact. Nikita Khrushchev's de-Stalinization program had failed and he had been removed from being head of the Soviet government in a bloodless coup. Leonid Brezhnev, a throwback to the Stalin era, was now in power. Just eighteen months before the Gates family arrival in Germany, Brezhnev had deployed 250,000 troops into Czechoslovakia to brutally crush the Prague Spring, denying citizens anything that resembled a democracy. The Cold War was just as intense as ever.

My assignment was to the 2d Battalion, 54th Mechanized Infantry Brigade, of the 4th ID. British forces were positioned to defend the northern half of West Germany's border, facing off completely with East German territory and Americans forces held the southern half. The East German border continued south until linking up with Czechoslovakia. Most of the 4th ID's area of responsibility was the Czech part. The entire border was very intense at the time. There were not any rounds exchanged, but there were a lot of small things that went on to cause continuous tension. Our cavalry units had a tough time, they were either always on the border or in training.

The reality was that if Moscow ever wanted to invade West Germany, that would have been the time to do it. Vietnam was the US military's priority for personnel and resources. The primary forces standing in the way of a Moscow-ordered invasion were the British and German militaries. But most important of all, the Soviet Union was still being led by tired old grey men who survived World War I, the Bolshevik Revolution, and

World War II. They understood and did not want to relive the horrors of war. Like all American Army commands in Europe, our battalion was operating at less than fifty percent strength. With the command operating at such low strength, we were well below the bottom level of being rated non-combat effective. As there were few captains available in the battalion, companies were being commanded by "green" lieutenants.

During my in-briefing, the battalion commander stated that with my background he needed me to serve as the intelligence officer. This was a captain's position. He explained, "We have no one else like you. We cannot stick a second lieutenant in there because it would kill him." I did not volunteer for the S–2 position, but I did not say no. The entire S–2 section consisted of just two people, a specialist on his first tour of duty, and me. The specialist was a good soldier who made it his personal quest to maintain our M577 mobile command post. The fact that this tracked vehicle operated well was a direct result of his continual preventive maintenance. He maintained the chassis, track, engine, radio, and every other piece of equipment. That specialist fully understood and never needed reminding that if that vehicle could not deploy and maneuver in the field we would be sitting in an iron rock. Unfortunately, few people in the battalion shared his view.

With the S–2 assignment came the opportunity to attend formal in-country Army schooling for operations, intelligence, personnel security, and physical security. Working with the command staff, with all the knowledge that came with it, was critical in my development. Equally important was learning with whom to facilitate work and how to communicate with senior staff officers, battalion and brigade commanders, and the division's general officers. Not having an officer for the S–2 position was a golden opportunity for me, but it was a challenge. Typically, the specialist and I worked long hours five days a week and part-time on the weekends. We did occasionally take a weekend off, but the duty officer always knew how to get in contact with us. At all times, at least one of us two was available to immediately return to the battalion.

It was an extraordinary experience to serve on the staff. In this position, communicating and coordinating with division G–2 on handling classified documents was a continual process. Most reports made there were sensitive so learning how to classify and control documents was critical. One primary responsibility was to analyze intelligence reports from brigade and division, and then apply the intelligence to battalion operations. Dealing with all those classified reports, both received from outside sources and created within our shop, was the most demanding part of the job. We had all kinds of contingency plans. I had some experience in handling documents and working with the division staff and the division G–2 from my previous tour in Vietnam. However, the volume of documents, along with serving as a derivative classification authority, was far beyond anything before. All classified documents had to be maintained correctly. Caring for and securing them was no problem. Self-recognition of a lack of knowledge had forced me to constantly research classification regulations and strictly follow the instructions. It took a lot of time, but there was no other responsible option.

Unfortunately, our battalion was having problems understanding this in most other areas of responsibility. Within less than a year, it would catch up to us. All commands had their annual inspector general (IG) inspections. The first year I was there, the battalion continued its tradition of failure. The common denominator of the battalion's problems was the commander. This non-combat experienced officer had the mistaken belief that if everyone liked him and their assignment, people would be happy and get along with each other, racial and drug problems would disappear, and such an environment of contentment would increase the percentage of reenlistments.

In a staff briefing, the battalion commander announced to us what was going to happen. In his "nice-guy commander" concept we were to report to duty at about nine o'clock in the morning, with the commencement of a continental breakfast. He then followed up by stating he would provide guidance on when to start training or whatever he wanted us to do for that day. At this briefing, all his company commanders and staff officers listened intently to everything he said. I was the only person who spoke up. "Colonel, with all due respect, I don't understand what in the hell you're talking about. Every person in this battalion should be doing company-leading PT at six o'clock in the morning. If you wait until nine o'clock, it will be ten o'clock before anyone is on the job. How are we going to maintain these trucks? How are we going to get training done?" The battalion commander verbally stumbled around trying to find an answer.

Because he was a lieutenant colonel, the captains and lieutenants thought he knew what he was talking about and accepted his instructions. It got worse. This battalion commander further instructed that NCOs were forbidden from entering troop living areas. He considered these rooms to be soldiers' private spaces into which supervisors were not authorized to enter. In all companies, access keys to the rooms were removed from the CQ's key rings.

Within six months of my informing the battalion commander that I did not know what he was talking about, we once again failed an IG inspection. Patience and confidence of the division and brigade commanders ceased to exist. The battalion commander was fired, and a battle-hardened lieutenant colonel took the reins of leadership. The next morning, the entire battalion found itself back in the Army and commencing PT at six in the morning. Just like in the Berlin Brigade under Colonel Weyand, the new battalion commander was doing all the exercises with us.

We immediately commenced hardcore training—starting with individual and small unit training in the garrison area or at field sites. What we were doing was not phony stuff. We trained the way we would fight. Unfortunately, we had a lot of catching up to do. The previous commander's debacle had resulted in the loss of perishable skills in even the most common soldier tasks, as well as in section operation tactics, techniques, and procedures. But persistence and strong leadership began paying off. We were back on the rise.

The new battalion commander's aggressiveness got us much-needed ammunition and spare parts, especially for our M114 Command and Reconnaissance Carrier vehicles. The M114s were fast, but piles of junk. They were assigned to Europe because they proved useless in Vietnam. They were mounted with 20mm guns that could not take out much of anything. The concept was good: it was small, had a low silhouette, and could swim. The M114 could outrun any other mechanized vehicle in the American and Soviet inventories when it was operational, and it would have to outrun the enemy as it was pretty much worthless for doing anything else. By equipment authorization standards, the M114s were assigned to the battalion commander, the operations sections, and the scout platoon. Of all the people in the command, these were the ones who needed reliable mobility, which the M114 did not provide.

Then came our new Assistant Division Commander, Brig. Gen. George S. Patton IV. Like his legendary father, Patton IV was a tough fighter and had earned a great reputation in Vietnam. One day we had an alert. Our vehicles were lined up in the motor pool and, except for one administrative problem, were ready to move out. Patton IV arrived and asked me, "Sergeant Gates, how come we are not moving?" I told him we had to get the keys from headquarters to unlock the gate. He said, "I will tell you what, how would you really get out if this was a real emergency?

Let us see you do it." Our scouts knew the way to take their vehicles through a fence was to strike the support poles with the tracks. At the pole, there was no spring in the attached fence, and there was no danger of the fence sliding up over the top of the vehicle and decapitating the gunners. Two vehicles side by side hit poles, creating plenty of room to allow the rest of the vehicles to follow. Like in the Hollywood classic about his father, Patton IV cheered us on as we passed by. That brief event was a major morale boost that stayed with us. On all further alerts, the keys were the first things to arrive at the motor pool.

Throughout much of my time in the 4th Division, we endured the same problems that plagued much of the Army in Europe. The two biggest problems affecting leadership were the lack of seasoned officers and NCOs, both in numbers and quality. Second and first lieutenants were only serving one year in each grade before being promoted. They did not have time to develop professionally. Had the OCS route been taken, it could have been feasible to have a twenty-year-old captain on our hands. Assigning lieutenants with little to no platoon leader and unit executive officer time was asking for disasters and NCO development was just as bad.

Our command was almost ineffective. A lot of the men who were in Germany at the time had been drafted into the Army. By the time a draftee had finished basic then advanced training, went to Vietnam, and left, there were only about six months remaining on their conscription. Taking away in-processing and in-country training and out-processing at the end of their tours, we were only able to work them for about five months. It just did not seem smart to send them to Europe after Vietnam. Most of them were very good soldiers, but they had been in a bad atmosphere and had the mindset, "What are you going to do to me, send me back to Vietnam? Hell, I have already been there." That stated, most of them worked hard and did their best to finish out their remaining months and received an honorable discharge. They had served their time in combat. They should have been assigned from Vietnam to stateside bases where they could get acclimated back into American society, have closer access to their families, and start looking for civilian employment.

Drugs were everywhere on the base and the ease in getting them was unbelievable. It was obvious that the drug market operated both on and off base. It was common belief that the bars downtown were the main suppliers. A joint German-American law enforcement operation could have helped reduce this problem. The German police considered this an American issue and was not much help.

Then came racism problems. The Department of the Army had identified the solution as awareness training and teaching soldiers the importance of respecting each other's diversity. Discipline was challenged. NCOs performing CQ duties were responsible for maintaining order in barracks during off-duty hours. Their authority and safety were being threatened by gangs that had developed. The battalion commander who forbade officers and NCOs from entering soldier quarters had exacerbated the problem. His successor removed this restriction on unit leadership, but the problems continued to get worse. Soldiers did not dare to go to the latrine alone for fear of getting attacked. Walking around outside had to be done on the buddy system.

Within the first year of my time at Bamberg, a soldier was convicted of a felony. Two guards were escorting him to Manheim Disciplinary Barracks by train. One of the escorts failed to remain alert, and the guard was killed with his own gun, and the prisoner escaped. The felon was subsequently caught and also convicted of murder.

Things really changed for the better the last year I was there. We had received a new brigade commander who ordered all doors removed from troop living areas. The CQs were then armed with loaded .45–caliber M1911A1 pistols. He allowed the diversity training to continue but added rules-of-engagement training for all soldiers. He wanted to make sure all troops understood that CQs and duty officers understood deadly force was authorized if life-threatening self-defense or defense-of-another situation existed. The brigade commander told all the officers and senior NCOs, "You are responsible. If I need to put you on latrine guard duty, then you are going to go on latrine guard." His aggressive leadership overcame most of the racial problems. The troops knew that if they tried to organize a protest, they would be arrested, court-martialed, convicted, and sent to jail in the disciplinary barracks. Racist undercurrents still existed, but behavior modification was achieved. Individual soldiers and unit leaders knew the brigade commander, supported by Patton IV, had their backs.

By 1972 the Vietnam War had slowed way down. As a result, commands in Germany started to get good, career soldiers with combat experience. Instead of sending draftees to us to finish their remaining months in service, we were sent fresh replacements straight out of AIT who would be with us for two and a half years. Spare parts for our vehicles and equipment began to arrive. The atmosphere in the unit totally changed, from making do with what we had to making things better. Now, all vehicles in the battalion were able to roll out of the motor pool in less than half an hour during alerts.

Eventually, we received the M113 armored personnel carrier-tracked vehicles that had previously been sent to Vietnam after the M114 fiasco. The fact that the Army attempted to replace the M113s with the M114s proved a lack of proper research, development, and field testing prior to deployment. A lot of people did not believe it, but the M113s were made to swim. On land and in water, we now had the mobility and reliability we needed.

The battalion personnel strength went up to about eighty percent, enough for the battalion to be combat-effective. We started getting more senior NCOs and company-grade officers. Captains, instead of first lieutenants with less than two years in service, were becoming company commanders. People in leadership positions were finally understanding what the All-Volunteer Army was and was not about.

After my two years of service as the S–2, a captain arrived at the battalion and was assigned to the intelligence office position. I would have been content working with him, but the new battalion commander had a different idea. We had developed a close professional relationship, and he appreciated the fact that the S–2 was working closely with his S–3 section. A captain served as the operations officer, a position that should have been held by a seasoned major. The captain was a good man, hard-working, intelligent, and eager to learn. The battalion commander assigned me to be the operations sergeant with the implied mission of mentoring the captain. It did not take long before the captain knew how to run an S–3 shop. Although our time together only lasted about six months, we had a relationship that should ideally develop between experienced NCOs and all young officers.

Since I had been on the master sergeant promotion list for the past two years, for the remaining six months of my tour of duty I was headquarters company first sergeant. The battalion commander had a philosophy that it was always better to learn the responsibilities of the next pay grade before putting on the rank. Once in the rank, the expectation was to already possess the knowledge of what the job entailed. Whether the rank is a senior NCO or officer, stumbling through on-the-job training was not an ideal situation—especially from the viewpoint of subordinates.

Shortly after becoming a first sergeant, a new commander was assigned to the headquarters company. Like other captains we had recently been receiving, he was a very competent officer. He was also a Ranger with Vietnam time. That was a lesson some soldiers had to learn on their own. Because the officer was black, a group of black soldiers in the company thought he was going to be their pal. The leader of the group came into the administration area one evening during the commander's open-door policy time. Previously, before someone was to see the commander through an open-door policy, they had to work through their chain of command first. The "All-Volunteer Army" changed that. He arrived, asked to see the commander, and I introduced him to the captain.

The first thing the soldier said to the commander was, "Hey bro, how you doing?" The captain jumped out of his chair, ordered the soldier to the position of attention, and stated, "When you come into this office and when you ever talk to me you will show all due respect." By the time the captain finished what he had to say, the soldier was trembling. The soldier made the mistake of thinking the matter was over when he complied with the captain's orders to get out of the office. But now it was my turn. That soldier was again standing at the position of attention and was told with no uncertainty that he would respect the captain. He was further informed that he and his friends would act like professional soldiers on and off duty or we were going to be having more of our one-way conversations. As with the captain, the soldier never stopped shaking the entire time I ripped into him.

Not far away was the rest of this soldier's circle of friends. He went back out and told them what had happened. I went back into the captain's office, closed the door, and apologized. Then the two of us started laughing about the soldier's sudden transformation from cool arrogance to trembling fear. Word of the encounter quickly spread throughout not just the company, but the entire brigade. Never again did the captain or I have any acts of insubordination or even racial comments within the command. The brigade commander had already cleaned up racism on the surface. Within our company, with just one encounter, that new captain cleaned up all remaining undercurrents. By the captain immediately taking aggressive action against such an act of disrespect he set the standard for his entire command. Even though it was a headquarters company, with senior officers over him, everyone knew there was a new marshal in town who was going to enforce the law and not play games.

Racism from any direction has no place in the world and certainly not in the United States military. It is one of the biggest cancers that eat away at unit cohesion and trust. Too many great Americans have fought and died to end racial inequality and abuse in American society. It must be attacked head-on when encountered, and accountability must be exercised on the perpetrators.

The US Army in Europe had turned the corner in both personnel and readiness. If not for the fact that my family and I were ready to return to the States, I would have enjoyed sticking around longer just to work with this captain. On the home front, that tour of duty in Germany was critical in bringing our family together. We got off to a rough start. Two weeks after arriving, the command had a forty-five-day exercise in Hohensfeldt. Margaret was on her own. Our vehicle had not yet arrived from the States. She had no transportation to get around the base.

In 1970, professionally managed spouse support organizations did not exist. Years later Margaret told me she almost said, "To Hell with this stuff, I am going to go back home." She did eventually meet other military wives and they helped her out. Our car finally arrived and we got a second one for Margaret. It did not take her long to find her way around the base and the entire city. Margaret became friends with a lot of Americans and

Germans. Memories of what she endured for those forty-five days turned out to be a good thing. Margaret was determined that other spouses would not endure such difficulties and those memories set the foundation for reforms that she diligently worked on as I moved up through the ranks.

Despite the rough start and heavy work schedule, our family built positive memories that have lasted our lifetime. We took the opportunity to travel throughout the continent. A lot could be accomplished with thirty vacation days a year. Military-sponsored rest and relaxation sites were located throughout Europe. These sites provided safe lodging and meals to service members who wished to enjoy themselves while learning about the surrounding area. With Bamberg already located inside Bavaria we had quick access to numerous castles. Oktoberfest was always fun for the entire family, and Switzerland was only half a day's drive away.

During our first year together in Germany we drove to Paris. There were hotels military could stay cheaper than any hotel back in the States. Our room had a nice balcony from which we could see the Eiffel Tower. The hotel was a short walk to the Seine River where the three of us looked at the historical buildings and bridges. Margaret loved art and adored the Louvre Museum. Where I only saw paint on a canvas, Margaret could decipher a painting and tell you what it meant. She spent an entire day in that place just looking at paintings. On this same trip, we drove through Spain, where we saw a bullfight and stayed at an American Air Force base for only fifteen dollars a night.

The same thing happened outdoors. European landscapes cannot be equaled anywhere else. While Margaret saw the beauty of the trees, grasslands, hills, and unique structures, I saw them from a military perspective. At the Black Forest while she saw the majestic trees, I was thinking how difficult it would be to maneuver mechanized forces through the pines. When we came to a river she looked at the beauty of the water and the surrounding countryside; I focused on the massive steel bridges that certainly would be a target in the next war.

The next year we took a trip to Italy and went to Rome. Unforgettable was the Pope coming out onto his balcony and blessing the people in the Vatican courtyard on Saint Peter's Square. Melissa started making the Sign of the Cross like the Catholics. We got her a cross that the Pope supposedly blessed. At the Coliseum while Margaret was looking at the architecture, I was thinking about the chariots and the Roman fighting tournaments. On all those trips we had an inexpensively good time.

While stationed in Germany, our second daughter, Lauren, joined our family. When Margaret went into labor, she had to be medevac'd by helicopter to Nuremberg. Being born in Germany created a lifelong problem for Lauren. She was issued a German birth certificate. Whenever my daughter needed a copy of her birth certificate, she had to go through a months-long process with the State Department.

Mine was a feeling of satisfaction in sharing all these experiences with Margaret. For all she had been through holding the family together while I was in Vietnam and now away from the family most of the week in Germany, she was finally able to enjoy the great things of life. We did not ever talk about the stuff we went through during our absences from each other. Margaret was the silent hero. Melissa was the dedicated trooper. The divorce rate in the military was high, but Margaret and I were two of a kind. We loved each other so much, and I worshipped the ground she walked on. I called her a lady with Southern grits. Margaret was always my calibration in life. She alerted me to my nightmares and brought me back to the present. Because I was such a hardcore leader, one time she told me, "Bill Gates, I love you, but sometimes you can be a pain in the butt." She was right.

Because we did not drink alcohol, we usually stayed home after a day's work. One evening, later in our tour, Margaret and I were in the NCO club with some friends. Nearby was a table with some loud male patrons expressing their judgment of women in the club. Margaret had gotten up from the table to go to the ladies' room. One of the individuals made an obnoxious sexually related comment about her. I got up and informed him she was my wife and asked him to make no further comments about her. He got up and stated he would say anything he wanted about her or anyone else. Now we had the entire club's attention. I found out later his current behavior was in keeping with the long-term problem other club patrons had been having with him. Neither I nor any other witnesses could say for certain if his subsequent actions were attempted intimidation typical of a bully, a bluff, or if he intended to make physical contact as he moved closer. If he thought I was going to back down and seek appeasement, he was mistaken. Within three seconds he was on the floor with my foot on his throat. Having witnessed the entire event, the club manager ran over to me and said, "Don't crush his throat." I replied, "I am not going to crush his throat. But he is going to respect my wife and every other woman in this establishment." The military police arrived and escorted him out. He did admit he was out of line and had started the confrontation. He had no choice as there was a room full of witnesses that night and countless times before. Later, I learned that after our encounter, he never bothered anyone else again. Why no one ever stood up to him before had no acceptable explanation.

A couple of weeks later, I was in the post exchange when a woman came up to me and said, "You are Sergeant Gates." After my acknowledgment, she said, "You took down that man for harassing your wife." I said that there was a minor confrontation, but nothing serious. With an appreciative response she said, "No, you stood up for your wife and all of us."

Photo 1. A young Sergeant Gates stands in an unknown drop zone wearing his parachute harness, drop bag in hand.

Photo 2. Command Sergeant Major Gates, while serving with the 2d Armored Division (Forward) in Garlstedt, West Germany

Photo 3. As Sergeant Major of the Army, Gates was a frequent visitor to field training exercises.

Photo 4. Keeping tabs on training standards, Gates would often visit Army Training Centers to see the latest technologies in use.

Photo 5. Gates is talking candidly with soldiers.

Photo 6. With the 25th Infantry Division, walking a dismounted training lane.

Photo 7. Always ready to put himself at the soldier's level, Gates is on a firing range listening to the instructions of an observer

Photo 8. Representing the American soldier, here Gates and Army Chief of Staff General Carl E. Vouno meet with a military dignitary of the People's Liberation Army of China.

Photo 9. Often traveling with the Army Vice Chief of Staff, here he and Lt. General Gordon E. Sullivan are talking with soldiers.

Photo 10. Sergeant Major of the Army Bill Gates visiting the troops during Operation Desert Storm.

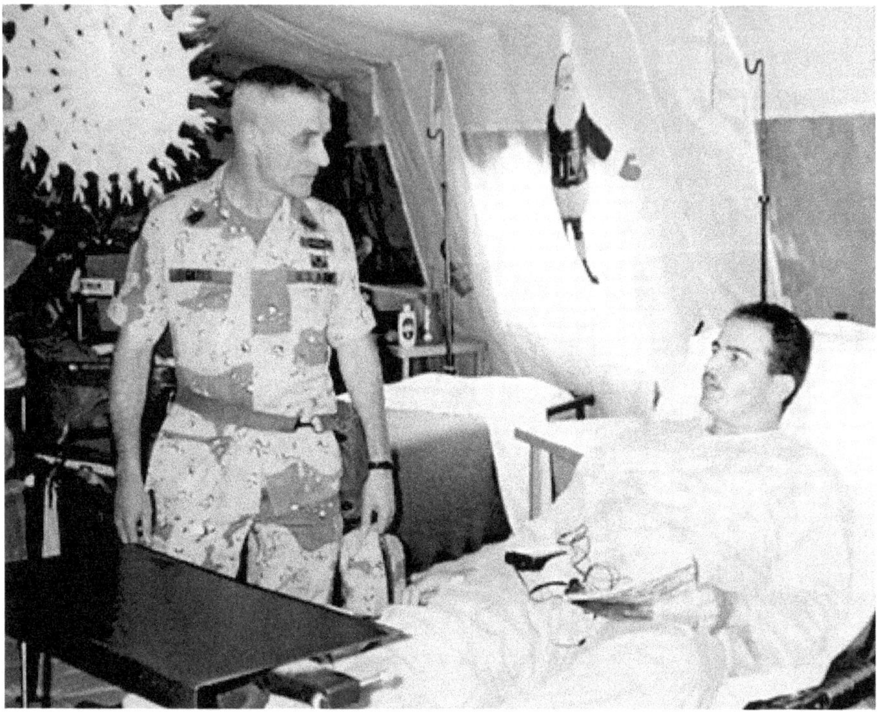

Photo 11. Pictured comforting a soldier at a field hospital during Operation Desert Storm.

Photo 12. Gates and Vouno pose with soldiers in front of a Blackhawk helicopter.

Photo 13. Listening intently to soldiers while in the shade of camouflage netting during Operation Desert Storm.

Photo 14. Post-retirement, he remained committed to developing soldiers and units. Here he is pictured visiting the Sergeants Major Academy at Fort Bliss, TX.

Photo 15. An expert trainer, Gates served as an advisor to the Army Research Institute at the Joint Readiness Training Center, Fort Polk, LA. Here he is seen in discussion with two role players.

Photo 16. Always with one more story to tell, Gates, with the then-current and four other former Sergeants Major of the Army, bestows recognition upon General Sullivan.

Photo 17. A frequent guest at Fort Polk, LA, Gates was speaking at Warrior Field during the 244th Army Birthday celebration on June 14, 2019.

CHAPTER 10
Be All You Can Be

Our family had a lot of mixed feelings when we finished our time in Bamberg in early 1973. The friends we made both on the American base and in the German community had become a special part of our lives. Civilians back in the States had to spend a lot of money for a dream vacation of traveling through Europe. What they could perhaps do once in a lifetime, we did at every opportunity available without stretching our budget.

When my family and I returned to the United States, there was a great sense of satisfaction. We had come closer together as a family. We felt like we had added to the community, both inside and outside the base perimeter. Duty-wise, I felt the hard work and dedication to my command and the troops within it had made a positive contribution to mission readiness and combat effectiveness. We had earned the honor of knowing we did the harder right, by each other and for those around us.

At the completion of an assignment, a soldier should always look back and take time for an honest self-evaluation. The critical question to be asked is, "Did I make a positive difference?" Accomplishments in the S–2, S–3, and as first sergeant, provided me with satisfaction that all paychecks received were earned. The battalion had come a long way in three years. Now, if the Soviets did want to launch an assault, we were ready and looking forward to the fight. We had no doubt that before the long fight was over, we would be driving our military vehicles through the streets of Moscow.

My next assignment was at the Mountain Ranger Camp in Dahlonega, Georgia. Located 125 miles northeast of Atlanta, Dahlonega sits among some of the most beautiful scenery in America. The town is host to North Georgia College. Dahlonega holds the distinction of having the first gold rush in America. The US Mint in Philadelphia operated a remote facility there from 1836 until the Confederacy took it over in 1861. Even today there are still people prospecting for gold. Occasionally, some pieces of gold are found. What used to be a precious metal industry in Dahlonega has turned into a tourist attraction.

The Mountain Ranger Camp was a satellite operation of the Ranger training program at Fort Benning. There is rugged terrain near Dahlonega. About three miles from the camp is the Tennessee Valley Divide. My assignment was to serve as chief instructor for the patrolling committee. The cadre was composed of mostly senior NCOs and officers of at least captain rank. There were even two majors. Every person on that staff was there because of their maturity, experience, and qualifications. Most captains would get promoted to major within a year of completing their assignments. Our executive officer was a lieutenant colonel. In time, the Mountain Ranger Camp became a battalion-level command of the Ranger Training Brigade.

As was the standard for a battalion-level command, the operations officer was a major who knew when to support us, and when to stay out of the way. All the planning for operations was done by the committee chief captains. The major focused on providing command assistance and coordinating external support, including helicopter, air, and whatever else was needed. He was able to consolidate and coordinate all the requests. It also worked out better for us to have a field grade officer requesting support from other commands.

The students lived in small wooden buildings with potbellied stoves, constructed during the Great Depression by the Civilian Conservation Corps (CCC). Throughout Georgia and the Carolinas, CCC workers built the highway that opened the region to the rest of America. Another regional project they completed was the Chattahoochee National Forest. The buildings had been built to serve as temporary lodging for the construction crews as they proceeded with their work.

The CCC was a Franklin D. Roosevelt New Deal project that employed young men ages eighteen through twenty-five. For their work, they received food, shelter, clothing, and thirty dollars each month. Twenty-five of those dollars had to be sent back to their families. It was a total win project. The men were provided with basic needs and learned skills that made them more competitive in the future job market. The work was meaningful to them, benefited the country, and employed over 3,000,000 people during its duration. Started in 1933, it was discontinued in 1942 after the beginning of World War II. Most of the young men transitioned into the armed forces. Having already lived under austere conditions and within a goal-oriented team structure while in the CCC, they were well-conditioned for military service. The barracks they left behind served us well forty years later.

Making use of the southern Appalachians, our job was to teach mountaineering. This included how to ascend and descend cliffs, build rope bridges and cross streams using them, and advanced rappelling. Not for the faint of heart was navigating a mountainside cliff where Ranger students had to free climb a part of that mountain without any special equipment to firmly secure them to the rock for about 300 meters.

In the patrolling part, we taught them how to conduct missions. They learned how to maximize the utility of the terrain for the purpose of the patrols. A good example is that we would teach them not to cross the topographical crests of hills. We taught them to look for animal trails. Most people do not realize that animals, especially deer, do not usually cross over peaks. They go around the side.

The area we mostly trained in belonged to the US Forest Service, with whom we had a great relationship. Most of our coordination with the forest service concerned letting their rangers know when we would be in the area. Having the final say they never refused us anything. They liked us because we did not tear things up, chop trees down, or leave garbage lying all over. Ranger students were not allowed to do that. Any time we fired blank ammunition, some civilian would collect the expended brass. We maintained strict control of our live rounds. Their picking up the expended brass was fine because it was almost impossible for us to find it all, especially by the night patrols. Whatever they received by selling the used brass we considered their earnings. The guys from the Forest Service would always let us know when they were going to put trout in the streams so the soldiers could fish for a meal. Inserting patrols by parachute or helicopter required us to reach out to the local farmers. This was not a problem either. The landowners also had a close relationship with the military. The fact that many of the students and cadre grew up on farms enhanced the relationships even more. The Georgia farmers mostly grew hay and raised cows. Some had small patches of peas, cotton, and tobacco. The aircraft pilots and jumpmasters ensured we were only parachuting into hayfields or cow pastures. The troops preferred the hayfields. They never learned to enjoy landing, stepping, or slipping on all those cow patties. Pulling in the chutes after landing in pastures always came with a manure collection.

Even with the farmers allowing us to use their land, parachuting into that terrain was not easy because the drop zones were very small. Ranger students hitting trees was acceptable because they had to learn. The positive relationship extended beyond the farmers and into the community. One day, I went into the town hardware store to find a piece of equipment. Unfortunately, I did not have enough money with me to buy it. I asked the store owner to hold on to it so I could come back on payday for the purchase. He asked me if I was with the Rangers. When he found out I was, he told me to just take it and he would look forward to seeing me at the end of the month. Standing by our agreement, he was one of the first people I saw after leaving the pay line.

That would not happen in the average city, then or today. In Dahlonega, we had the best of relations with the civilian population. We also never had any problems with the college students. They were just as much pro-American as anyone on the base. If someone did try to start something, local citizens would have brought a quick end to it. We were not just stationed in that community, we were part of it. In turn, whenever their schools or other community organizations needed support, we were there.

While I was stationed at Dahlonega, the Army finally decided to start developing an NCO education system, and an Advanced NCO Course was established at Fort Benning, Georgia. I was among the first to attend. The course lasted about eight weeks and was centered around first sergeant, operations sergeant, and master sergeant duties. Although still a sergeant first class while attending the course, I had a head start by having already worked all those positions. Not resting on my prior experience, I fully applied myself and was a distinguished honor graduate. As the Army continued to develop NCO education, this course eventually became a course for Sergeants First Class. That was a good decision.

In 1973, the Army was again thinning out its ranks. Once more, officers who did not keep up with the requirements of their commission were becoming sergeants—so the NCO promotion system was jammed up again. Eventually, it would free up and there would no longer be extended delays in promotions.

Starting in 1974, there came a series of moves and reassignments that would affect our family for years. As often happens, the owner wanted to sell the house we were living in. We moved to the other side of town to a beautiful house on a small farm. The farm was about ten to fifteen acres. We rented about three of them. There, my wife and kids got really attached to the owner's mother who lived right next door to us.

Orders promoting me to master sergeant finally arrived and the promotion came with reassignment to the Third Ranger Company at Fort Benning. My new job was to serve as a first sergeant with administrative responsibility for all cadres and students of the company. Sometimes the three phases of patrolling, mountaineering, and the Florida phase would be happening concurrently. It was interesting and challenging to provide administrative support to three missions going on hundreds of miles apart. Challenging did not mean impossible; it meant I had to always stay in the game and not fall behind on my responsibilities.

A game changer for me was the introduction of Standard Installation and Division Personnel Reporting Systems (SIDPERS). This was the beginning of the Army's computerized tracking of all personnel actions. The old paper DA Form 1 Morning Report was complicated enough, but filling out those color-coded letter-sized cards for each personnel action was a nightmare. The cards would be consolidated at the battalion and sent to off-base headquarters. One wrong pencil coding entry and the entire card was kicked back, and the process started over. When it was refined, and all the problems worked out of it, SIDPERS was a step in the right direction. The Army assigned Fort Benning to be the pilot program. We never figured out if this was because of our motto, "Infantry Leads the Way," or senior Pentagon leadership figured if a bunch of infantrymen can succeed in working SIDPERS, so can the rest of the Army. The pilot program succeeded, and the rest of the Army was brought on board.

One of the biggest things I addressed while at Dahlonega was confronting problems facing the rest of the Army. The Home of the US Infantry, Fort Benning, like all major Army installations, was ground zero for a series of issues. Because of the effects of Vietnam, the NCO and officer corps were broken. The clear-cut line that separates NCO and officer responsibilities had faded. Lieutenant platoon leaders and captain company commanders were making decisions that should have been left to staff sergeant squad leaders and sergeant first class platoon sergeants. First sergeants were staying closer to their offices overseeing the work of their administrative and training staff. Command sergeant majors were becoming enlisted advisors to commanders at the battalion and above levels. The people who allowed this to happen were the generals and the command sergeant majors. Maybe this was because when the generals were young officers they found it easier to do the job themselves than have it done again. That may be an excuse, but it is not a justifiable reason for failing to enforce standards and let subordinates professionally develop.

Fortunately, I did not experience that as much as many other supervisors. One reason was that I wouldn't allow it to happen. Second, if the NCO was not worth his pay, then we would do something with him other than simply keep him in the Army. I refused to transfer my problems to other commands. Perhaps the NCO in that category was never trained by professionals. A lot had to train themselves and had to pull stuff off by themselves. Perhaps they never had an ultimate role model like I did with Sergeant Victor Franco.

The same applies to incompetent officers. Some people are too lazy to professionally develop, too arrogant to accept their own shortfalls, or cannot progress because of their own limitations. Too often, the preferred solution became giving these people nice evaluations and passing them on to other commands. That was a post-Vietnam-related problem that also continued to get worse. The times created the problem; lack of professional leadership and accountability allowed it to flourish.

The Army had to go through a lot of trials and tribulations during the 1970s and early '80s. Figuring out how to recruit, train, and equip volunteers was not the main problem. Once the oaths of enlistment were completed, they were soldiers. The problem was adapting to Secretary of the Army (SecArmy) Bo Callaway's visions of an All-Volunteer Army while concurrently trying to rebuild and train the force.

Once in position, Callaway should have just assigned the mission of developing an All-Volunteer Army to soldiers and allowed them to figure out the "how" part within a specified timeline. Just as officers and NCOs proved we could make something as complicated as the original SIDPERS program work, we should have been left alone to make the other changes. Instead, Callaway and his team of whiz kids micromanaged the Army and forced their untested ideas into policy.

Simultaneously, the US Marine Corps was filling its ranks without lowering standards. Instead of trying to play appeasement, the Marine

Corps maintained themselves as "The Few and the Proud." They proceeded into the 1970s maintaining discipline, camaraderie, and professionalism. If not micromanaged from the E Ring of the Pentagon, every command in the Army could have done the same.

On a personal level, my dedication to duty and determination to work hard was recognized twice over. In 1976 I was selected for promotion to staff sergeant major and received orders to attend the Sergeants Major Academy at Fort Bliss, Texas. While at the Academy I learned my next assignment was once again to be a first sergeant. I was not disappointed because it was going to be in the soon-to-be activated Ranger Battalion.

CHAPTER 11
Ranger Company First Sergeant

Not since World War II had there been a full-sized Ranger Battalion. General Order 127, issued by US Army Forces Command (FORSCOM), activated the 1st Ranger Battalion with an effective date of January 31, 1974. Army Chief of Staff General Abrams had recognized the need for a highly trained battalion-sized reaction force that could quickly mobilize, not in a few days or hours, but immediately upon notification.

More than perhaps any other soldier in the Army, Abrams knew what a fast-moving battalion could accomplish. Without lead time, in what became known as World War II's Battle of the Bulge, he had been a battalion commander in a corps that had been ordered to turn ninety degrees north and link up with the encircled 101st Airborne at Bastogne. The 101st was doing a spectacular job in holding the road network, but they needed reinforcements and armor as soon as possible. Time was of the essence.

As the lead armored battalion commander in the assault, Abrams had been instructed to follow the road network moving north. Nearing Bastogne, realizing strict compliance with the order would result in bypassing the 101st perimeter to the west, Abrams and the lead infantry battalion commander took it upon themselves to shift advancement to the northeast. The speed of the American assault toward Bastogne and the willingness of the battalion commanders to adapt to a fluid combat environment formed a critical milestone in ending the Third Reich.

Abrams knew from the existing US Army where to draw the troops necessary to fulfill his vision. From World War II he was aware of the accomplishments of the Ranger battalions. From his days as senior commander of allied forces in Vietnam, he knew what Rangers could accomplish behind enemy lines. He also knew he already had enough school-trained Rangers spread throughout Army ranks that in constructing such a battalion, very strict soldier selection criteria could be exercised.

First assigned were senior officers and NCOs. Sergeant Major Glen E. Morrell was selected as the battalion sergeant major. Morrell and I did not need to spend time figuring each other out because that had been accomplished years earlier. Glen would later precede me as Sergeant Major of the Army. We believed that in the 1st Ranger Battalion, NCOs would lead and train the troops. From the start, there was no misunderstanding between Glen, the commanders, and me that my time within the command would be limited.

The selection of personnel continued through June. Completion of Ranger School was not enough for acceptance into the battalion. That only started the consideration process. We were looking for the best of the best. A proven record, dedication to duty, firm self-discipline, and a high level of motivation were the standards that determined who would be allowed into the command. Anyone whose records reflected a failure to maintain those standards in previous commands was immediately disqualified. Also, personal knowledge on the part of officers or senior NCOs considering an applicant's poor performance resulted in immediate disqualification. The volume of applications received allowed us to be very selective.

As selectees started arriving at Fort Benning, they immediately went into training. By the end of June, we were at full strength. On July 1, 1974, we held the activation ceremony at our newly assigned Fort Stewart, Georgia base. Ours was not simply a static formation ceremony followed by a pass and review. We set out to make a statement. As many as the parade field could support, soldiers of our activating battalion came in by parachute.

Once on the ground, the parachutists assembled into their respective company formations, and the battalion marched forward to the viewing stand where General Abrams awaited. His presence, backed by his accomplishments and personality record, was always an inspiration. General Patton had called him "the best tank commander in the entire Third Army."

We had to develop administration, training, and supply from scratch to get the battalion and the subordinate companies operational. We did have proper facilities for messing, supply, vehicle operations, and all other unit functions. Except for messing operations, everything else was in a continual receiving process.

From their previous units, we received personnel files on all our soldiers. Consolidating those records into a central filing cabinet was the only easy part. SIDPERS was still in its infancy and caused me to be distracted from other things I should have been doing as a first sergeant. Our solution to running the orderly room was a soldier of that unique breed who is as comfortable jumping out of an airplane in full combat gear as at a desk surrounded by paperwork. It was wonderful to find a good person like that. There was still an administrative load, but one I could complete in a fraction of a day rather than having the day totally consumed. Across the Army, there were and still are first sergeants who allow themselves to be trapped in an orderly room. I was not one of them.

The thoroughness of that soldier's work allowed me to simply become the reviewer before anything was sent to the commander for approval. The soldier even requested a second set of eyes on what he had created. He understood that administrative processing is like writing a document. The worst proofreader of any document is the person who developed it. At least three lessons to be learned came from that soldier. First, individuals should always be recognized for skills they possess outside their assigned specialty. Although he was not primarily working as a Ranger, he greatly contributed to the command.

The second lesson concerns the command operational mission. To get the administration completed, we were taking him "out of hide" and not being able to use him in all our missions. This made his squad one person short and less effective. Even though squad members saw it as their contribution to command success, it still took away from their team being fully functional.

The third lesson is for all leaders, all units, and all time. Working soldiers outside their assigned specialties while doing critical missions elsewhere is harmful to their professional development. Their advancements into the NCO ranks are based upon how well they perform within their specialty, not some other function. Being denied individual and team training opportunities corrodes their proficiency. All skills are perishable.

It is wrong to expect soldiers to study off-duty to maintain their skills. They should not have to make up for lost time while their peers relax. Individual study by manuals never compensates for hands-on experience and team training. There will always be an administrative workload waiting to be done. Not all of it must be completed immediately. Whenever possible, the soldier was sent out to do Ranger training. In his case, the combination of Ranger and administrative skills came together well. He went on to earn a commission through OCS, eventually retiring as an officer. Not all soldiers will push themselves as hard. Some will push themselves too hard. It is the responsibility of their leaders to ensure the right balance is maintained.

In supply operations, being a newly activated battalion, we had to request and stand by to receive all the items of authorized equipment. In the end, this worked out well. Since we were Abrams' brainchild, about ninety-nine percent of our requisitions were completed within a month. Everything was right off the factory assembly line. Probably not, since World War II was an entire battalion outfitted with new M151A1 general purpose (Jeep) vehicles. Previously, I had never seen a new Jeep and thought they were all supposed to be old and subject to breakdowns from wear and tear.

For special missions and field exercises, we often needed additional equipment or vehicles temporarily. A formal request sent up through an infantry brigade to the division usually produced positive results whenever our requests could be supported. Sometimes, we did not need to go through formal chain-of-command routes. Picking up a phone and calling another first sergeant was an option. C Company, 1st Ranger Battalion, was always very good at extending support.

Networking at the NCO club was also effective. One evening, I mentioned the need for fifty ammo pouches to another command's sergeant major. The next morning, those pouches were delivered to my office. Everything was hand-receipted and returned in as good condition as we received them. Likewise, when other commands needed equipment, we were there to help.

Rarely, but sometimes, our formal and informal systems could not get us all the equipment we needed for a training mission. We did not get upset because we knew we had tried, and no one was holding back support from us. Instead, we recognized operating with an equipment shortage as a training opportunity. The senior NCOs and officers had served in Vietnamese jungles, plains, or mountains. They knew how to fight with what they had available.

Having outstanding administration and supply sections is important and contributes to mission readiness. Training would ultimately determine battlefield success. My assignment as A Company first sergeant was largely due to the time I spent at the Ranger development schools. Instead of building individual Rangers and teams, we were building companies and a battalion. Fortunately, my previous first sergeant experiences had already taken care of what could have been a learning curve.

The units our soldiers had served in since Ranger School had varied in both types and quality. To ensure everyone was proficient in common tasks and Ranger-specific skills, we initiated an aggressive training program that started with the basics. First came common tasks and standard infantry skills. For those who had maintained skill proficiency, this was considered refresher training. For those who had lost proficiencies, we saw it as "rust removal." Everyone was being brought up to speed together.

A senior officer was there with us, but normally, the NCOs worked with officers to plan exercises. No matter who it was, we protected our lieutenant in the field because we did not want him to get hurt physically or career-wise. NCOs did individual training. Platoon leaders and sergeants would meet with the commander and executive officer before we went into the field. The officers participated in some, but more in a monitoring role. They also wanted the NCOs to lead the operations as much as possible. The battalion S–3 training sections were there for oversight and support when needed. The good thing about them was that they allowed the companies to develop and execute their training. Sergeant Major Morrell made sure battalion micromanagement did not exist.

That kind of balance works exceptionally well within any unit. All the leaders got involved with all the operations. Rather than just saying to "move over there," for example, the commander coordinated with the battalion and shared everything with the lieutenants and NCOs. The lieutenants were his representatives with the platoons, working with the platoon sergeants. Even as first sergeant, I did not make decisions before coordinating and liaising with the NCOs. We had a fantastic system in place to prepare for and execute missions. The soldiers would see the unity between the officers and NCOs. Witnessing harmony between the leaders supported a teamwork mindset.

There was one decision I made and enforced without consulting anyone. Throughout the Army, soldiers were required to paint rocks lining walkways to their barracks and administration areas. The colors represent branch colors, such as infantry blue, artillery red, and so on. Painting rocks is one of the dumbest wastes of time for soldiers. It does nothing to enhance the appearance of the unit area. Once the process begins, it never stops as the paint fades or chips off the rocks. I eliminated this wasteful mandate. Instead, we used the time to train soldiers.

The billeting and administrative areas assigned to us were already available. In the 1960s and '70s, the Army had moved away from the old-World War II wooden and Korean-era Quonset Hut types of barracks to cement/brick-and-mortar type structures. These three-story facilities were well constructed and designed for at least two companies to live in one building. On the third floor of our building was a transportation company. The transportation soldiers thought it was amusing to throw their used cigarette butts out their windows into our police call area. That stopped when my A Company Rangers threw one of the transportation company soldiers and his discarded cigarette butt into a dumpster. I learned about it when Morrell stormed into my office stating, "Gates, what the Hell are you doing?" We got the transportation soldier out of the dumpster. After he went on his way, we told our soldiers never to do that again. If not for the requirement on our part to warn our soldiers against a recurrence, our orders would have been unnecessary because never again did a transportation soldier toss any kind of rubbish into our area.

Working closely with platoon sergeants and squad leaders was critical. The mission was to get them all to move in the same direction and to work among themselves. In addition to training and mentoring subordinates, senior NCOs also had the same responsibility toward officers.

Weapons qualifications received a big push. We had M16 rifles, M203 grenade launchers, and both M60 and .50–caliber machine guns. Everybody in the command had to be qualified with all weapons. Normally, commands are allowed only enough bullets per year for weapons zeroing and one attempt at basic qualification. That does nothing to help improve individual weapons skills. The thought of a Ranger qualifying only as a marksman is bizarre, at best. We did not even need sharp shooters. All Rangers should qualify as experts. In combat operations, the split second it takes to shoot the first round into the heart or head may determine who lives and who dies. The engagement of multiple targets requires every shot to be a kill.

The use of artillery was a big thing. In ideal conditions, artillery lieutenants and NCOs serve as forward observers and reconnaissance sergeants and deploy with infantry units. Ranger teams do not operate in ideal conditions. They were going to be out front on their own. Every command member was trained and evaluated in calling in and adjusting artillery fire. The two most critical things about calling in fire missions are knowing how to read a map and operate a radio. Failure at either could be disastrous.

In various forms, Army aviation was always there for us. We had a close relationship with the aviation battalion and trained as one team. Helicopter crews love to operate their aircraft under various conditions and missions. In the 1st Ranger Battalion, training was conducted on using attack helicopters to provide fire support and to clear an area of the ground. As in Vietnam, helicopters were one of our primary means of insertion transportation. We trained hard on three maneuvers: tactical, fast roping, and parachuting. Tactical was primarily sitting on the floor of a Huey helicopter, then jumping out as it touched the ground or hovered slightly above it. Tactical could also include sling loads for additional equipment and supplies to be transported under the helicopters. Rappelling was good for deploying onto rooftops or other places where circumstances prohibited helicopters from getting too close. Consisting of multiple motions, rappelling was not good for speed insertions when tactical options were available.

We also trained on helicopter touchdowns. We would receive information that helicopters would arrive at a specific time. Before the helicopters came in, teams were already on the ground to secure the area so that the helicopters were protected inside our perimeter. Higher command never needed to tell us how to secure the area; we were just told of the critical areas that needed to be defended. It was up to our experience and leadership skills to know the security configuration positioning of defenders.

As the helicopter approached, we deployed smoke grenades, allowing the flight crew to recognize how the wind was blowing. Radio communications were always maintained. Sometimes, we would simulate that the area had turned hostile when the helicopter was about to land. We would have to wave off the pilots and shift to an alternate location. Parachute jumps were at least once every sixty to ninety days. We could jump every thirty days, depending on the situation and how much equipment was available.

Being co-located and having a good relationship with the aviation battalion, getting helicopters was seldom an issue. Scheduling C–130 Air Force planes took more coordination. These missions were not just troops jumping out of an aircraft but included equipment drops. Our purpose in life was to be placed behind enemy lines. Knowing how to go in with all the necessary equipment to ensure victory was essential. The element of surprise only lasts a short time.

Two unique aspects of our training compared to normal infantry units were how we deployed them into the field training sites and what we did upon arrival. We would come in fast and move out with fire and maneuver engagement on pre-positioned targets and mock-enemy positions. A lot of work went into setting up these exercises. The most important concern was ensuring that weapons fired from simultaneously deploying Ranger teams did not have the remotest chance of crossing into each other's maneuver spaces. The preparation teams did their jobs well, and we never came close to a stray round endangering anyone.

The individual squads and fire team leaders came together like a synchronized machine. They would have to determine the size of the unit needed to accomplish the mission. If they only needed a platoon, we would not send them to a company, but they would stay in communication with it. If the platoon got into more than they anticipated, they had quick reinforcements. Even though it was a training environment, we always had another unit ready to deploy. These scenario escalations had already been planned out by the battalion training staff based on our Recondo experiences in Vietnam. For good reason, the information needed for the exercise to progress was not prematurely provided to the deploying teams. They had to "earn it." This was "train the way you fight" at its finest.

Combat support and medical support was also standing by for simulated injuries. Even the base hospital was on alert to receive mock evacuees if so requested. Our field exercises were a training opportunity for them as well. If someone was injured, we had all the medical and airlift support to attend to them immediately and get them to the hospital within minutes.

Without effective supervision, the finest administration, supply system, and training are useless. The worst example is Union General George McClellan in the American Civil War. He mastered organizing and training of the Army of the Potomac. Had he known how to lead it in combat, the Civil War would have been over in its second year at the Battle of Antietam. Instead, his inability to exercise aggressive leadership caused the war to last three more bloody years, costing hundreds of thousands of casualties on both sides.

We had no such leadership deficit within the battalion or the company level. All commanders were good soldiers, providing supervision and guidance as necessary. They were willing to stand back at the appropriate times to allow the NCOs to do their jobs. When our company commander was promoted to major, he told me his success was because I had made him look good. It was quite a compliment. We had teamed together, along with the lieutenants, the NCOs, and the junior enlisted. Having received daily guidance from the battalion commander, he kept his subordinate leaders informed and created a professional environment where we could succeed.

Eighty percent of my time was spent with the troops. Already having served as a first sergeant in Germany, I did not have to make a mental transition from platoon to company leadership. I was combining first sergeant skills with Ranger operations. Bringing two separate experiences into one role was more enjoyable than difficult. Whenever the soldiers went to the field, I would go out with them, except when unavoidable circumstances mandated remaining in the administrative area. As soon as what caused the "stay in the rear" mandate was over, I moved forward to be with the troops. Soldiers always perform best when their leaders are present. This is not just because they were being supervised. The platoon sergeants can handle that. Soldiers appreciated senior leaders sharing the same challenges and hardships that they did. They respected leadership by example.

Senior leader presence is even more critical in high-risk military professions—Rangers certainly being one of them. The benefits of tasks must be balanced with the risks involved. In the rear, reviewing of a risk analysis matrix is appropriate during the preparation phase. In forward operations, matrixes are useless. The leader on the ground, seeing firsthand what is going on, listening to subordinate concerns, receiving weather and situation reports, sometimes backed by nothing more than gut instinct, formulates the best risk assessments. A seasoned leader can do this very quickly. Avoiding challenging training just because of potential risks is wrong. To do that means we would never jump out of planes into combat zones. It means not taking foolish or unnecessary risks, thinking we are invincible.

Rangers must fight with every team member being fully effective. Bad stuff will happen if not. It should never happen because leadership is sloppy and does not conduct full assessments before mission commencement and continuing those assessments during mission progression. If anything goes wrong, the first sergeant and commander should be immediately available. Things going wrong are not necessarily bad unless there is loss of life, serious bodily harm, or destruction of critical equipment. Under almost all circumstances, things going bad are part of the learning process. These are targets of opportunity that afford leaders a chance to help subordinates develop solutions. Things that go wrong in training exercises are destined to go wrong in combat. Problem-solving is part of individual, team, and unit development.

Soldiers are most influenced by squad leaders, platoon sergeants, and the first sergeant above them. Platoon sergeants are the ones in positions of ultimate trust. Squad leaders are junior NCOs in the development stage. Platoon sergeants must also be the direct link to the first sergeant and company officers. Platoon sergeants must always act professionally, understand tactics, know the importance of equipment maintenance, and establish a firm relationship with each soldier as an individual. Some soldiers needed little guidance and motivation, while others required close oversight and leadership. The platoon sergeant should be able to operate on a sliding scale as the individual and situation requires. If a soldier has a problem and the squad leader cannot resolve it, the platoon sergeant should be the first person engaged.

As first sergeant of my company, I had four primary roles: First was to set the example for my soldiers and my unit, including the company commander. Second was to take care of my soldiers. This included knowing each soldier, whatever problems that soldier might have, and trying to solve them. Third was to coordinate with the battalion to ensure all personnel and materials necessary to maintain combat readiness were received. Fourth was ensuring the command continually conducted effective and meaningful training in preparation for combat operations.

Listening to and monitoring soldiers were critical responsibilities of all battalion leaders, especially platoon and first sergeants. Sometimes, a soldier would not openly express a problem. That is where monitoring and observation came into play. A serious change of behavior or attitude meant an external influence or issue affected the soldier. The leader needed to talk to them, discover the source of the change, and take mitigating actions.

In the 1st Ranger Battalion, standards were maintained in the field and off duty, whether on base or in the civilian community. Throughout the battalion, the Rangers knew if they committed violations of the Uniform Code of Military Justice (UCMJ), they would be disciplined. The knowledge of accountability, backed by the fact that we had hand-picked the best soldiers for the battalion, resulted in our not having many discipline problems. Sometimes, not related to discipline, a few soldiers had issues meeting the demands of being in the 1st Ranger Battalion. If it were a training problem, they would receive extra mentoring. It is understandable that some people do not learn at the same pace as others. It was not uncommon that someone who took a little longer to learn retained the lessons longer. Extra mentoring of a well-meaning soldier produced a good return on investment.

If the problems were motivation or attitude-related, the soldier was provided with three counseling sessions. Either the Ranger straightened up and pulled his weight, or he was sent packing for having proven not to be up to battalion standards. They were not bad soldiers, just not up to the pace we were setting. The problems we encountered were never so serious as to remove poor performers from the Army with a less-than-honorable discharge. We would have had no hesitation in doing so if a soldier warranted such action or if there had been a discipline problem.

Our success at forming the battalion, developing administration, supply, and training operations, enhancing the development of Rangers as individuals and teams, and fine-tuning our own leadership skills was frequently tested through immediate response training exercises. The battalion was always on alert, meaning it had to have one company available to respond immediately. The companies took turns fulfilling this alert requirement. During this period, only emergency or medical leave was authorized. The first sergeants worked together in lending troops between companies to ensure the alert company could deploy at full mission strength. During these alert periods, we trained hard because teams were at one hundred percent operational strength. Immediate response exercises were welcomed, as the assignments were unpredictable and challenging.

Our favorite activity was a no-notice mission requiring us to fly to Fort Irwin, California to rescue a general officer whom OPFOR had captured. The base was the predecessor to what would become the National Training Center. Meanwhile, the OPFOR had been assigned the mission of ensuring the general remained in their custody or to simulate killing him if his rescue could not be prevented.

No members of the OPFOR were told exactly when or how the rescue operation would be launched against them. They were probably expecting ground troops or mechanized infantry charging the perimeter to execute the rescue. Either way, getting to the vicinity meant vehicles transporting the rescue teams to a nearby assembly point, disembarking, and going over land. In the desert at night, automobile headlights could be seen for forty miles. The OPFOR had every reason to believe they would receive plenty of indications and warnings when the assault would be launched. Perhaps even the OPFOR officers and NCOs were unaware of the existence of the newly formed Ranger Battalion. They were about to get an education.

We made a night jump far enough from the OPFOR compound not to be seen or heard. It was common for aircraft to come and go from the main base, so our plane did not attract special attention. While conducting our operation, the Air Force crew went into the main base to refuel and waited for our pick-up call. We had a good parachute landing, quickly organized ourselves, and commenced movement to contact. The noise and light discipline were excellent. Undetected, we were able to get into position for a good reconnaissance of the perimeter. The compound had a 360-degree fence, and guards were properly positioned. The guards missed the only indication and warning they could have alerted to, and that was that all the coyotes in the area became quiet as we passed by.

Now we had to figure out how to get in there without losing a bunch of soldiers and rescue the general before he was killed. In our mission brief, we were advised of the building where the general was being held. Every rescue idea we came up with had flaws. One soldier said he knew how to get in, win the mock gunfight, and rescue the general before the OPFOR could react. His plan was simple. Those are the best kind, and we went with it.

Through our communications links, we contacted the base fire department and reported a fire inside the compound. Caller identification on telephones was still years away, so the fire department had no reason to question the source of the call. With only one blacktop road in the compound, identifying the route of travel for the fire engine was easy. While patrolling the road, we found an excellent depression in which to position ourselves and block the fire engine from view when we stopped. The emergency message was sent, and soon, a fire engine came into sight. An unexpected bonus was that an ambulance was closely behind. We stopped both. The emergency crews were initially annoyed by the false alarm. They must also have seen the Hollywood war classic *The Dirty Dozen* because they quickly accepted our plan. With most of the team inside the ambulance and a few more on the fire engine, what now became a Trojan Horse operation commenced.

We raced forward with lights flashing and sirens screaming. OPFOR soldiers opened the gates, and we went straight into the compound. Without losing a single Ranger, we neutralized OPFOR and rescued the general. There was an airstrip nearby and our now refueled plane landed for extraction. On the flight back, we learned of having a casualty; one of our Rangers had broken his ankle but never said anything about it. He said he hadn't wanted to mess up the operation.

With my pending promotion to sergeant major and the longevity of my time at Fort Stewart coming to an end, I knew a new assignment was in the near future. How near I did not know until the 24th ID commanding general called me to his office. Upon my arrival, the division's command sergeant major explained the meeting's nature. Both he and the general were tired of spending money to send their sergeants to the NCO Academy at Fort Benning. They wanted their own academy to focus training on the 24th ID mission to cut the travel expenses. My new role was to be the Commandant of the 24th ID NCO Academy.

The meeting with the general went well. He outlined his expectations and assured me that all the support necessary would be provided. Another small CCC compound away from the main base was available, and all primary division staff officers were ordered to support my mission.

The compound needed upgrading, which the engineers provided. G-3 operations worked closely with US Army Training and Doctrine Command (TRADOC) and other NCO academies, allowing me to obtain lesson plans and course material. As the engineers did their magic, G-4 logistics ordered all the desks, beds, wall lockers, and other life support items. Meanwhile, G-1 administration was recruiting potential staff and instructors. None were accepted until they passed an interview process led by me. Even the post exchange got in by providing a mobile trailer to sell hygiene, administrative, and other small items. With all the support, we quickly became functional. All the lesson plans were examined and adapted to the division's mission and the instructors were trained and rehearsed, including professional standards that they were required to maintain. Next came the TRADOC certification, which we passed with flying colors. With the opening of our doors, training commenced.

The Academy was going well when orders for my next assignment arrived. For the following two years I was to serve as Chief Instructor at the Virginia Military Institute Reserve Officers' Training Corps (ROTC) Detachment. The relaxed pace gave Margaret, Melissa, Laura, and me more time to spend together. Considering what would await us in 1980, those days in Southwestern Virginia were very valuable.

CHAPTER 12
Garlstedt

My third tour of duty in Germany, the second for my family, came in 1980. The new assignment was for Garlstedt, near Osterholz-Scharmbeck, twenty-one miles south of the Army's Bremerhaven port. Most US military logistics that came into West Germany passed through that port.

My family liked returning to Germany because of the positive experiences from the previous tour. This time, we were in the northern sector, allowing quick access to Denmark, Belgium, and Holland. The cities and smaller communities in those coastal countries, along with Lower Saxony and Germany, had some of the most unique structures in Europe. Even though we were getting older, we still enjoyed visiting places like Copenhagen (the home of Hans Christian Andersen) and Hamelin (the German community from which came the Pied Piper story). Any assignment for a soldier and the family is what they wish to make of it. We made the most.

Germany had changed a lot since my days in the Berlin Brigade. Berlin itself had grown out of the rubble of World War II and was a major metropolis despite being surrounded on all sides by East Germany. West Germany had become a major 20th-century industrial power, and East Germany was an economic wreck. The Soviets were still very much in control of East Germany but no longer testing and teasing British and American forces as they had done for the first thirty years after World War II.

Operating in the central and southern parts of Germany, the United States developed its doctrine on the concept that if war erupted, the main Soviet thrust into West Germany would come through the Fulda Gap. Despite high command rhetoric and troop acceptance throughout the ranks, military history and geography dispelled this scenario. Had the Soviets invaded Western Europe, their main thrust would have been across the Northern German Plain. This is the same terrain used by military forces since the days of antiquity. It is outstanding tank country and ideal for expedited movement of forces.

The Northern Plain offered the best route to seize German, Belgian, and Dutch ports. Either capturing those ports or making them unusable for Allied shipments would have been a strategic gain for the Soviets. That's where the 2d Armored Division (Forward) came in. It was to ensure Bremerhaven remained free of attacks from hostile forces and to repel threats while the British military conducted ground and air operations against the Warsaw Pact. My new assignment was to serve as the command sergeant major of the subordinate 2d Battalion, 50th Infantry.

Commands isolated from mainstream US Army have a habit of going one of two ways, depending upon the professionalism of their officers and NCOs. With strong leaders, units will maintain discipline and mission focus. They can even become better than units on larger bases because they do not have the distractions. On the other hand, isolated units can slide down until sloppy operations are accepted and tolerated. The 2d Battalion had taken the downward spiral.

There were only two or three Airborne Rangers in the battalion. They were the battalion's exception in understanding that discipline was the foundation of professionalism and a critical part of camaraderie. Enough years had passed since Vietnam and most junior enlisted lieutenants and captains had not served in Vietnam. The seniors had failed to remember lessons of combat or had done their tours inside base perimeters. They were about to experience a very fast culture change.

My initial walk through the motor pool was on the first Thursday after my arrival. Vehicles were parked haphazardly, and rubbish was all over the place. The problem was so bad that the only thing the perimeter fence was good for was catching newspapers, food wrappers, paper cups, and other trash that should have been removed during daily police calls.

A thief or adversary trying to break into the motor pool by cutting through the wire mesh fence would never have been spotted by the on-duty guard. That made me realize that guard duty was also not being taken seriously by leadership. Maintaining fields of observation is one of the most fundamental principles of both stationary posts and walking patrols. This rubbish should never have escaped the attention and corrective action of the duty officer, sergeant of the guard, and NCOs working inside that motor pool every day.

Upon returning to my office, I summoned all the first sergeants. Observations were shared, and my instructions were calmly given that within twenty-four hours, that motor pool was going to be cleaned up. The first sergeants assured me that everything would be corrected by the next day. I went to the motor pool on Friday afternoon and found nothing had been accomplished. Remaining in the motor pool, I summoned all first sergeants to my location. They had been offered a chance to do it our way, which meant they could have followed through with my previous instructions and engaged their subordinates in resolving the problems. Now they were going to do it my way. As we walked through the motor pool, I addressed every deficiency, from the fence line to the vehicle line, to rubbish in the vehicles, and to the janitoring of the maintenance building. Tools and oily rags have a place for accountability, personal safety, and fire prevention. Lying around haphazardly was not it.

Their failure to correct the problems as they promised made it obvious that the first sergeants were not able to supervise subordinates in cleaning up the motor pool as they had been instructed. It was explained that because of that assessment, they needed both additional leadership and task-specific training. Then came the order they were not expecting—all NCOs, from junior sergeant to myself, would report for duty on Saturday. We were going to clean up that motor pool to regulation standard. Junior enlisted would not be included. We were going to perform "Train the Trainer" and, in keeping with an old classic song, "I Did it My Way."

When we finished that Saturday, the motor pool looked sharp. The following Monday produced a surprise. The base cross-country running course went around the outside perimeter of the motor pool. Being division-forward, we were commanded by a one-star general who did his morning run on that course. My first detailed conversation with Brigadier General James E. Armstrong was when he walked through my office door. He stated, "I have never seen that motor pool so clean and organized as I did this morning." I tried to deflect the praise by saying, "It was the first sergeant's initiative." That response fell flat on the floor between us. Armstrong replied, "Those first sergeants have been here for months. It was cleaned up the first weekend after you arrived. That is not a coincidence." Immediately upon returning to his office, Armstrong informed his subordinate commanders that each one of them was to emulate what had been done in our motor pool.

The lack of professionalism concerning the motor pool grounds and maintenance building was an indication and warning of another problem. Vehicle maintenance became my next area of focus. The battalion owned Jeeps, five-quarter pickup trucks called Gamma Goats, armored personnel carriers, two-and-a-half ton and five ton cargo trucks, and other mechanized equipment. These vehicles had been getting on for years. Inspection of company maintenance records and individual operator technical inspection worksheets for their assigned vehicles revealed neither operator nor unit-level maintenance were being conducted. Vehicle logbooks were not being maintained.

My inventory of the vehicle's prescribed load list revealed the failure of both unit maintenance and supply programs. Shortages should have been recorded and on requisition. The same rule applied to vehicle replacement parts. There were always funding difficulties and delays in getting parts. Claiming it was an unsolvable problem had been accepted by the command. Properly worked, those issues could have been resolved. The same battalion and company motor sergeants who let their work areas become overrun by rubbish were not doing their jobs with the equipment. I was even more disgusted with these specific motor sergeants because, generally, throughout the Army, their peers had not engaged in complacency.

Inspecting company driver records and requiring soldiers to show their operator permits revealed they were operating vehicles without certification. Driver training, road certification, and maintaining qualification records took time. Instead of doing the harder right, for the supervisors, it was easier just to toss a soldier a set of keys and let him drive. That problem was not just a motor sergeant issue, it involved every supervisor who knew subordinates did not have licenses yet allowed them to operate vehicles. It also involved senior leaders who did not provide proper oversight and do spot checks on their own.

If any soldier without proper certification had ever gotten into an accident with a German vehicle on the highway, the first thing the investigating police officer would demand to see was the soldier's driver's license. Once a German attorney realized his client had been injured in an accident with an uncertified US Army driver, the American government would have been in an unwinnable lawsuit. Even if the accident was not the fault of the soldier, the German attorney would be able to argue that the soldier lacked the experience to properly operate the vehicle and take evasive action to avoid the collision.

Both the certification of the drivers and maintenance of the vehicles played into soldier confidence or lack thereof. If they had to deploy into combat, neither they nor their equipment would be up to the challenge. When soldiers knew something real was happening, not an alert, and their vehicles had not been taken care of and would not get them where they needed to go, it was a serious degradation of their morale.

It took a little longer to fix the vehicle issues than it did to clean up the motor pool. Training was the first step involving driver qualification and proper conduct of maintenance. Thorough record-keeping was mandated. Soldiers came to understand the expression, "If it is not documented, it did not happen." World War II General Mark W. Clark was right in saying, "Soldiers do what the commander checks." That works all the way down the line. Ultimately, it worked out so well that we never had problems with motor pools or maintenance after that.

Any time after that when I had a driver hand me the vehicle logbook, the maintenance and usage records were always found to be up to date. The knowledge that I was going to be checking the records resulted in subordinate NCOs doing their checks. At first, it was because they wanted to stay out of trouble with me. Soon, the reason became that it was the right thing to do.

Reflecting the "lead by example" standard set by Colonel Weyland years earlier in the Berlin Brigade, I did morning physical fitness training with the subordinate companies. Soon each of the companies knew I would be with them at least once a week. They just did not know which day, so they had better have it planned out and do it right every time.

Still in my second week in the command, after completing PT, I cut through the barracks area while returning to battalion headquarters for a uniform change. My shortcut preceded a police call by an hour and revealed chicken bones lying all over the ground close to the billets. The evening before, soldiers had been munching on fried chicken and then throwing the bones out their windows. In the morning meeting with the first sergeants, I informed them that they needed to "have the bones leave the unit and never come back. They need to go into the trash, not on the ground outside the windows."

By the time of the meeting with the first sergeants, the bones had already been policed up, but the message was clear. Food garbage lying on the ground all night attracts insects and rodents. Mice have fleas and lice. The problem was resolved for the next two days, but on the third day it was back. The troops had been allowed to toss the bones for so long that it became second nature. The first sergeants and platoon sergeants had not been following up and enforcing compliance. They had not learned from the previous weekend's motor pool detail that if our way did not work, it was going to be my way.

Each subordinate company was ordered to post a 24/7 "chicken bone patrol." While on duty, the guard was required to walk around armed with an M16 rifle, a full web belt, and a rucksack. To complete this uniform requirement, CQs were required to open the arms room and issue weapons for every shift change. For safety, we did not issue ammunition. The rules governing the use of deadly force did not include stopping chicken bones from being thrown out of windows.

On the fourth day, I was walking to the dining hall when a soldier was marching around outside with his weapon and rucksack. He asked for permission to speak to me. With permission granted, he proceeded. "Sergeant Major, if you let us stop this patrol, you will never see another chicken bone that was tossed out a window." I thanked him and told him I would take his word as his bond. I informed the first sergeants of the conversation that the detail was over and issued the warning that if the chicken bones reappeared, the detail was back on. There never was another chicken bone on the ground.

The companies started maintaining their surroundings fantastically. They kept the grass mowed. Daily room inspections improved not just appearances but also sanitation standards. A morale boost for the battalion troops was General Armstrong publicly bragging about our motor pool and living areas. The battalion commander was pleased and extended to me his appreciation of the accomplishment.

From the most junior enlisted soldiers to the battalion commander, they were recognized because they were good. That's something every soldier and every person takes satisfaction in hearing. The grumbling of the NCOs for having to come in on their day off and continual reposting of the "chicken bone patrol" was replaced with pride and confidence, which are two of the greatest force multipliers to exist within a combat command.

About two weeks after my arrival, a day was spent inspecting each of the arm's rooms, starting with the company headquarters. Security and CQ checks during off-duty hours were not a problem. The rooms had functioning alarm systems. The weapons were properly secured but not organized in a way to support quick issues while maintaining accountability.

We had the standard array of firearms, including M1911A1 pistols, M16 rifles, some with M203 grenade launchers, as well as M60s and M2 .50-caliber machine guns. I personally took apart and put together randomly selected weapons. The revelation was that the only time the weapons were cleaned was once after range qualification. There were no follow-up cleanings to remove dust and dissolve carbon.

Arms rooms must be organized, and weapons must be maintained to expedite effective deployment in the event of an emergency. Alerts were conducted to determine a unit's ability to quickly suit up, move out from the bases (or as the enemy calls them—prime target zones), and be ready to engage adversaries in the minimal amount of time necessary. Deficiencies identified during those alerts should have resulted in corrective actions. Fumbling around arms rooms and deploying with dirty weapons reflected a command that did not take alerts or their combat mission seriously.

My inspections were not "gotcha" visits. Each one was a training opportunity, in which each armorer and the supervisor received directions on how the arms rooms needed to be laid out. To their credit, by then they had picked up on the "our way versus my way" concept. The companies immediately went to work and fixed the problems. The follow-up inspections revealed that within a week the rooms were reorganized, and all weapons were clean.

While the arms rooms were being fixed, attention was turned to soldier weapons qualifications. Small arms qualifications were not a problem. M60 and .50-caliber machine gun qualifications were at the bare minimum. As already emphasized, it was not common sense that only the gunner could operate the machine gun. If something happened to the one person who knew how to operate a piece of machinery, then that machine could not be used. There had to be four or five backup operators. We were not going to be issued enough ammunition for everyone to become machine gun qualified. Disassembling and reassembling the weapons, knowing all the components, cleaning the weapons, learning malfunction procedures, getting comfortable with the sights, and practicing dry fire do not take ammunition. We were successful in getting enough rounds of ammunition so that every soldier could fire a few rounds in addition to certifying the primary gunners.

The motor pool situation, the chicken bones, and the arms rooms proved battalion discipline had been totally lax. Professional standards, as well as regulations and orders, had come to be considered optional. Walking by a problem and pretending it did not exist had become the norm. That kind of attitude is like cancer. The lack of professional soldiering goes beyond duty hours and affects off-duty behavior, including appearance, camaraderie, and professional conduct on and off post.

The communication sections were the welcome exceptions. The battalion communication sergeant took his job seriously and worked closely with his company counterparts. The only thing he lacked and had been asking for was the opportunity to do more unit-wide training. This was an easy fix. Under the umbrella of military occupational specialty qualification, we started doing radio training. We used the "crawl, walk, run" strategy of training. In the "crawl" stage, radios were mounted on the vehicles, and we started communicating inside the motor pool. The "walk" stage had radio teams sent out to different parts of the base to do communication exercises. In the "run" stage the teams drove around field sites and switched team composition so that the same people were not always together.

This training produced several byproducts. First was unit camaraderie because people not normally working together were now performing joint small-group training. Only four people could fit into a Jeep, and it was impossible not to verbally communicate with each other. Second, other skills qualifications like map reading were included with those communications exercises. Third, fuel was being pumped through the vehicle engines which, along with operator maintenance, improved their condition. Fourth, the troops appreciated doing realistic training rather than waiting for a stagnant day to pass.

Within a month, we had the NCO Corps running well, and across the board, they were fulfilling their responsibilities in their assigned sections. Things stayed fixed. We saw improvements even during the PT sessions where everyone showed up in the same uniform, and when they entered morning formations looking professional. Their uniforms were sharp, their boots were well-shined, and they had proper haircuts. It was even reflected in their off-duty behavior. Off-post discipline problems became a rare exception, and soldiers looked out for each other.

A few years earlier, the Army had started issuing Common Task Training (CTT) and MOS manuals for soldiers. The books were perfectly sized to fit into a field jacket pocket. These skills qualification books were a compilation of lesson plans. For someone to teach a class, all the information they needed was there. This was hip-pocket training at its finest. Soldiers hate the reality of "hurry up and wait." Supervisors could just reach into a pocket and conduct on-the-spot training.

My leaning on the battalion S–3 training section resulted in books being ordered and issued to all soldiers. There was no reason why every soldier did not have a copy immediately available. There came a realization among battalion soldiers that if they got into an even brief conversation with me, they probably would have to prove they had a CTT or MOS manual with them.

Hip-pocket training quickly progressed to team, squad, platoon, and company training. Junior NCOs were required to do much of the instruction. This forced them to study the subjects thoroughly, learn how to organize training plans, and become comfortable with providing blocks of instruction. Senior NCOs and I were present not only to make sure the instruction was correct but to add a little more pressure. We were preparing them for career challenges that were in their future. Developing them to accept and overcome pressure was critical in building their self-confidence.

All this should already have been in place. The battalion operations section should have been pushing training and skills enhancement. Training should have been greatly influenced by reports coming in from the intelligence section. The supply section should have been making sure all required equipment was present and functional. The battalion executive officer was responsible for bringing the staff sections together. By designation, he was the command's maintenance officer. The battalion commander's roles were to supervise the executive officer and the company commanders, oversee the officer development program, and ensure the battalion was ready to immediately "move, shoot, and communicate" in support of the division-forward commander's mission.

None of this should have been overwhelming at the time. If done correctly, leadership is rewarding. The biggest reward is realizing soldiers have confidence. Building that confidence includes making sure they have the right training and equipment to do their jobs. Failure to do so undermines soldiers' confidence and trust in their seniors. Once lost, those are two things that are difficult to replace, especially for the ones who were in charge when the failures occurred.

Mentoring and caring for soldiers, individual and team training, and ensuring equipment is ready to go were always the responsibilities of NCOs. Senior NCOs, like officers, should give subordinate leaders room to operate and grow themselves. Guidance should be frequent, oversight continual, and accountability enforced when necessary. There was not a single requirement or standard I placed on the battalion's NCO Corps that was not already covered by Army regulations. Except for the battalion and company communications sergeants, the rest of the NCOs had given up their authority in favor of having a relaxing time. They were comfortable being in an isolated environment certain neither they nor the equipment they were responsible for would ever be deployed into combat. The fact that they were assigned to protect the number one American military seaport on the Northern German Plain did not register with them. Had I arrived at the battalion and stayed in my office while the command staff and company-level leadership floundered, then I would quickly have become part of the problem.

Backed by the battalion commander, years of military experience including combat and mentoring from great leaders, forbade me from accepting complacency. With rank and position does not come privilege but the responsibility to fulfill requirements. The Army NCO Corps proclaimed itself to be "The Backbone of the Army." Once we corrected ourselves, the NCOs of that battalion became the backbone. With the strengthening of that backbone, the entire battalion developed into a well-functioning body. After a couple of months, my job began decreasing from being a prime motivator with an aggressive attitude where my directions were blatantly ignored to being a mentor. Like an aircraft that reached flight altitude, I was able to go into cruising mode while maintaining direction and speed.

This did not last long. Just months into my position as the battalion's top soldier, I was pulled up to division forward. General Armstrong liked what he witnessed in the battalion, especially the end of complacency. Using the 2d Battalion, 50th Infantry as an example of what could be accomplished, Armstrong had been successful in forcing the other subordinate commands to improve. He did not want them to improve just because he mandated it to be so, but rather because his subordinates were proudly fulfilling their responsibilities.

Armstrong wanted the command to be built up from within—specifically through the NCO Corps. In our initial discussion, he stated I was to be the enforcer who would spend ninety percent of my time with the units and ten percent doing paperwork. Our mutual agreement on that point resulted in the exchange of pledges. Mine was to accomplish the mission he assigned. His pledge was to provide full backing and top cover, knowing full well some toes were going to be stepped on.

There was a cultural change in going from battalion to division forward. The focus of a battalion is to have troops and equipment ready to accomplish missions designated by the senior command. Battalions have a specific specialty, such as infantry, armor, or artillery. If done properly, it was where the commander and the command sergeant major could have the greatest impact on the effectiveness, professionalism, and long-term development of the companies and the soldiers within.

2d Armored Division (AD) (Forward) consisted of its headquarters, six battalions (two mechanized infantry, one armor, one field artillery, one engineer, one support), and one company-sized troop of cavalry. As the division forward commander, Armstrong's mission was to look outward while trusting subordinate commanders to do their job well. On that subject, President Ronald Reagan had a great philosophy, "Trust, but verify." For Armstrong, verification was my responsibility.

Because of the work done in the 2d Battalion, all the commands knew I would be inspecting their motor pools, supply and arms rooms, training records, and facilities. They were given a couple of weeks to prepare before I started making my rounds. When the verification visits began, major problems did not exist. There are always minor things that require calibration adjustments, even among the finest units. Those were opportunities for mentoring.

As the senior American officer in the Lower Saxony region of West Germany, Armstrong had several other responsibilities that were consuming his time. He was the community commander. Normally, a community for a two- or one-star general would be the local base and a nearby training area. Our community stretched from the East German border to Holland and from the North Sea to the Hartz Mountains. For the American units in the region, all but Helmstedt Support Detachment, which was part of the Berlin Brigade, drew their logistic support from us. The US Army base at Bremerhaven was also within Armstrong's region of community responsibility.

The division forward resulted from a concept that did not work effectively. In 1976, "Brigade 75" had been developed to protect the Bremerhaven port area inside the British sector. The British Army of the Rhine was an outstanding command, but in the event of a Warsaw Pact invasion, its mission to engage the enemy to the east would have been impeded by protecting Bremerhaven. The manning of Brigade 75 was done by rotating soldiers every 179 days out of the Fort Hood-based 2d AD.

The concept proved to be flawed in several ways. The continual rotation of troops meant no continuity and a marginally effective working relationship with the British and German military in the region. As soon as troops arrived on the ground, they were short-timers, with short-timers' attitudes. Families were left behind in Texas, which for the many married soldiers translated to off-duty party time. With no consolidated troop areas, the brigade was spread out. Maintenance of equipment was viewed as doing only what was necessary to get through the 179 days. Every one of these flaws could have been identified by thorough analysis when the concept was in its earliest planning phase. Recalling the result of two-year draftees coming back from Vietnam who were sent to Germany to finish their military service should have been a red flag for the 179-day rotation idea.

Immediately upon execution, the Brigade 75 concept proved itself in need of revision. The result was a division-forward command led by a one-star general. The German government stepped up and offered real estate for a permanent base within Bundeswehr's Garlstedt Training Area. Construction of the entire base was done by Germans, and funding was shared by the American and German governments.

The base was completed in 1978. German-built meant it was to last. Our base at Garlstedt was no exception as it was the most modern installation in Germany. Everything was there: troop billets, family housing, a small medical facility, a mini post exchange with a snack bar, a theater, a gymnasium, motor pools, and whatever else we required.

Being built within the Garlstedt Training Range afforded us immediate access to maneuver space for tanks, artillery, and mechanized troops. Russian artillery would not have been able to reach either Garlstedt or Bremerhaven. Unless we were hit first by an airstrike, which was unlikely for a small base, being collocated with that training area allowed us the opportunity to quickly disperse our fighting equipment and troops in the event of a surprise attack on Germany.

A major responsibility of the command was developing and maintaining a close relationship with the British and German militaries. For the commanding general and me, this was probably more pleasure than work. Two world wars had proven how well the British and Americans could fight side by side. By having completed the British Ranger Course helped in the development of my mutual bond with their troops. As when I went through British Army Ranger training I found working with RSMs an enjoyable learning experience. In their commands, they were the most aggressive and knowledgeable of all the soldiers. They were always out front and knew what needed to be accomplished in preparation for both current and future missions. Never was any of their leadership authority and responsibility surrendered to their officers. The RSMs were a positive influence in the development of my own style of leadership as I progressed through the sergeant major ranks.

Working with the Germans required more effort. Both the British and the Americans had more in common with the Germans than with any other continental European nation. Their ties to the British went back to the House of Hanover and Braunschweig, which were both located in our regional community. Joint exercises at the local level were often conducted. One thing that became evident during these exercises was that the German Leopard tank was the best in the world and would remain that way until the Americans deployed the Abrams.

A nationwide annual event that had a major impact on us was the REturn of FORces to Germany (REFORGER) war games. In 1967, with Vietnam at its height, President Johnson had decided to pull two divisions out of Germany. The Soviet threat was far from over. Trust in NATO, especially the West German and British military capabilities, provided President Johnson with some room to maneuver. Confirming that the US military would be able to quickly return with full combat readiness resulted in the REFORGER strategy.

In 1969, annual exercises commenced, involving several NATO commands. Included were bringing in division-sized forces from the United States. REFORGER was more than providing a "show of force" to the Warsaw Pact. Exercised and stressed to determine the point of breaking, our logistics troops had to keep up with the movement of troops and convoys starting near France and working their way across central and southern Germany. After the annual exercises were over, the maneuver commands returned to their home bases. Then the review process began.

The REFORGER war games had been scaled back from the early 1970s but were still important to ensure readiness. When equipment for the exercise initially started coming into Bremerhaven, we were responsible for its security. As the units started arriving from the States, we turned the responsibilities over to their commanders and commenced our field deployment operations. In the event of a build-up to combat operations, this was exactly how the progression would unfold.

REFORGER also increased the activity of the Soviet Military Liaison Mission (SMLM). Throughout Germany, SMLM had agents studying American, British, and German military operations all the time. They were not allowed on our bases, but almost everywhere else was open to them. There were treaty-authorized spies dating back to the original division of Germany after World War II. Driving vehicles with bumper stickers validating their authority to observe, their missions were overt. The Russians were proud they had the opportunity to do that and would brag about it. Occasionally, we would grab one of their team members in places they were not authorized to be and turn him over to the authorities. This was not a one-way legalized spy operation. Based in Potsdam and operating throughout East Germany was the United States Military Liaison Mission.

The greater problem that required our constant attention was terrorism. After Andreas Baader and Ulrike Meinhof committed suicide in prison, their gang morphed into the Red Army Faction (RAF). In August 1981, making use of almost nonexistent entry control at Ramstein's main vehicle gate, RAF leader Christian Klar was able to drive an explosive-laden van onto the base and park it outside the US Air Force Europe headquarters building. The building was destroyed, but because the bomb detonated an hour before Air Force personnel reported for work a three-star general and his entire staff were spared.

Had such an attack been tried at Garlstedt, we might have lost an entry gate and the duty guards. That was as far as the terrorists would have advanced onto the base and they would have lost their assault team in the process. Professional terrorists always conduct extensive reconnaissance before organizing their attacks. The entry control security we had at Garlstedt was so tight that RAF would have gone off to look for a softer target. Armstrong and I routinely received intelligence reports from our G–2 and German police. These were beneficial, but not our primary justification in making modifications and enhancements to our security. As Vietnam combat veterans, we also relied on our instincts and knew if we saw a weakness, then so could the enemy. We often acted when something just did not feel right. Years later, this would be called "random antiterrorism measures." We were not so literate and just called it "leadership."

Winters turned our attention to another threat—road safety. Throughout Germany, one of the greatest dangers was "black ice." Frequently, by the time a vehicle driver realized the danger, it was too late. Soldiers residing in barracks were not a concern as they were collocated with their work areas and support facilities. Walking was their primary means of transportation. The danger was for soldiers living with their families in base housing.

In discussion with Armstrong, we came to a simple solution: to transport married soldiers between their quarters and the command; we would alternate between two-and-a-half-ton trucks and buses depending upon the severity of the risk. The unit alert system was used to inform soldiers when road conditions were unsafe for cars and that transport would be provided. We even had heaters installed within the truck cargo areas to prevent frostbite on the worst days. Because everyone lived on the base, there was no need to drive into the community during dangerous weather. The result was that not a vehicle incident occurred, nor was a single person injured.

Just how strongly Armstrong supported my work was proven over an incident that started in the consolidated dining facility. Often, on Saturdays and Sundays, I would go in and check on how well the troops were being fed. Without officers and senior sergeants likely to show up during weekends, mess sergeants would cut back on portions served. Enhancing their dining facility budgets at the expense of soldiers' diets was wrong.

One Sunday I arrived and went to the salad bar. The tomatoes were rotten and passed over by the troops. This was causing two leadership problems. First was health, and the second was that serving rotten food undermined soldiers' confidence in their leadership. I ordered the mess sergeant over to the salad bar and told him to eat one of the tomatoes. The tomatoes were sent to the garbage can. Soldiers in the immediate area overheard the conversation. That evening and into the next day the story spread throughout the command. Troop morale had a boost. If I was not already, through that incident I became their sergeant major.

The mess sergeant looked foolish and took it personally and his feelings were hurt. Monday afternoon I received a telephone call from the IG stating he wanted to see me. Looking back, I should have just responded that he was welcome to come to the command section anytime he wished. Instead, I was courteous and informed the general's aide I was going to the IG's office.

The mess sergeant had made a complaint about my ordering him to eat rotten tomatoes. This was the first time I ever heard of a complaint filed by a mess sergeant who was told to eat his own food. The IG immediately tried to take the high ground and accused me of issuing an illegal order. Rather than listen to the details, the IG was more interested in flexing his lieutenant colonel brass against my stripes. My stand was the problem with that mess sergeant serving substandard rations on the weekends would already have been solved if the IG had been doing his job and visited the mess hall on weekends. The IG and I got into an argument.

Losing both the argument and control of his emotions, the IG screamed, "I'm going to have you court-martialed." Those words had barely cleared his mouth when the office door burst wide open. Having been advised by his aide that I had been summoned to the IG's office, Armstrong dispatched himself and was standing outside the door as the conversation degraded. The general walked up to the lieutenant colonel and stated, "You called my sergeant major in here without telling me about it. This whole discussion is over. You are relieved. Start getting yourself ready to go home."

The IG was fired on the spot and subsequently transferred out of the command. The troops knew they had a commanding general and a sergeant major who would fight for them. I had a boss who was as good as his word and who had my back. Perhaps the ultimate leadership lesson was leaders must be willing to consume the same food they serve to the troops. The mess sergeant certainly got that lesson. There never was a rotten tomato or substandard meal served again in that command—at least while General Armstrong and I were there.

Working with Armstrong was a pleasure. He had so many responsibilities that he needed someone he could trust to look out for him at the soldier level. That ninety percent of the time spent with the troops and ten percent of the office ratio we set up worked well. It also allowed me to work with the subordinate command sergeants major and first sergeants, in identifying training objectives. I spent a lot of time in the field, depending on the operation. Because some troops thought I was snooping, they made an extra effort to do it right. The truth of the matter was I preferred to be with the soldiers than back in the office pushing paper. We were not just building units; we were building soldiers.

CHAPTER 13
Third Infantry Division Command Sergeant Major

My family would have been content if I served the entire three years at Garlstedt. This was not meant to be. In 1983, Major General Howard G. Crowell, Jr. succeeded Major General Fred K. Mahaffey as commander of the 3d ID. Simultaneously, the division's command sergeant major was also wrapping up his tour of duty in Germany. Rotation of both the division's senior officer and NCO at the same time, especially in an overseas command, can create both difficulties and targets of opportunity. Added to this mix was the simultaneous rotation of several brigade commanders.

General Crowell was an accomplished officer, having earned both the Ranger Tab and Senior Parachutist Badge. Gunfighting engagements in Vietnam had earned him the Combat Infantry Badge. Rather than reaching back to the United States for his senior enlisted soldier, he looked inside Germany. The outgoing division command sergeant major had brought to his attention the in-country presence of someone with two aggressive Vietnam combat tours, extensive Ranger experience, and a proven track record in building training programs and academies who was now serving his third German tour of duty. My initial meeting with Crowell proved we would be a complementary match, possessing similar leadership styles. Working with him and the division would be an honor.

In World War I, the 3d ID remained solid and counterattacked against a major German offensive in the Second Battle of the Marne. In World War II, as part of OPERATION Torch, it came ashore at Morocco, defeating the Vichy French in 1942. The following year under the leadership of Major General Lucien Truscott, the division led the fight into Sicily. After taking Palermo, it was Truscott's "Hail Mary" maneuvers that were critical in American troops outracing the British to Messina. Barely taking a breath, the 3d ID came ashore at Salerno and commenced fighting its way up the Italian Peninsula. The division further proved itself in the Korean War. The division did not see duty in Vietnam but rather remained in West Germany as a deterrent against the Warsaw Pact.

Until the yet-to-happen collapse of the Soviet Union and East German governments, the United States Army maintained two corps in West Germany, responsible for defending the center-south region of the country. In the center, the V Corps was headquartered at Frankfurt. On its southern flank was the VII Corps, headquartered at Stuttgart. Further south, on the VII Corps' right flank, and sharing the responsibility of defending southern Germany's Bavarian region, was the II German Corps. As part of the VII Corps, the 3d ID was headquartered at Würzburg.

In 1961, the unexpected construction of the Berlin Wall resulted in a reinforced company of the 3d ID being missioned to convoy duty north from Bavaria to West Berlin. This mission could have gone one of three ways: the reinforced company could have been blocked and forced to turn back, resulting in another Berlin Airlift; they could have been captured by East German forces; or a firefight engagement could have occurred, which would have annihilated the command while starting World War III. With all NATO forces on standby, the company arrived in West Berlin without incident. Two confirmations resulted: allied ground access to West Berlin was assured and the construction of the Berlin Wall would remain a limited-scope objective on the part of East Germany and the Soviet Union.

With democracy driving its government and capitalism driving its economy, West Germany quickly developed into one of the most successful nations in the world. For both defensive and economic reasons, allied military bases were spread throughout the country. Should hostilities have erupted with the Warsaw Pact, our subordinates would not be in one location and vulnerable to a single consolidated or nuclear attack. Even though the West Germans had constructed the finest road networks in the world, getting our brigades out of the garrison and into their field sites was a lot smoother without having them stacked upon each other.

1st Brigade was garrisoned at Schweinfurt, 2d Brigade at Kitzingen, and 3d Brigade at Aschaffenburg. The spread of those bases throughout the country also brought economic boosts. Civilian employment, the purchase of German construction materials and goods to operate those bases, and service member off-duty spending were critical in rebuilding local economies.

We were still years away from electronic communications such as e-mail and video teleconferencing. Between commands, documents traveled by courier, and conversations were generally conducted by telephone. If handled properly, the disadvantage of division and brigade leadership not having immediate face-to-face accessibility to each other was also a benefit. Brigade commanders and sergeants major had a lot of latitude in developing the subordinate commands and maintaining their bases.

The first order of business for General Crowell and I was being briefed by division staff. The second was to visit our subordinate commands to begin forming our own assessments. Combining knowledge gained from the staff and line soldiers, Crowell and I went to a private conference to examine the division's mission statement. We came up with some initial minor adjustments. From that, we built our command discussion points. We were in such close synchronization with each other that there was no contradiction in what was being said to subordinate leaders.

From the adjusted mission statement, Crowell ordered all his subordinate commanders to reexamine their respective Mission Essential Task Lists. Division headquarters went first, followed by brigades, then battalions, then companies. No subordinate command was to have a declared task that did not support the higher command's mission. Next came involving all soldiers in this process. It was typical for soldiers not to know their division mission statement or any of the interrelated documents. The 3d ID was no exception. Every soldier was expected to be a combat warrior. Command staff organized the missions but were not the ones who pulled the triggers. It was irresponsible for any leader not to ensure that all soldiers understood and remembered the mission of their command. One important role for leaders was to ensure their soldiers were aware of their importance and were provided with the knowledge to do their mission in the most complete manner possible.

It was not hard to resolve the awareness problem once the subordinate command sergeants major understood I was serious. Explaining the mission in morning formations and posting it on bulletin boards created awareness. Questioning during guard mounts and oral promotion board evaluations drove the message home. Senior NCOs knowing the division command sergeant major was going to ask the question when talking to soldiers during unit visits served as further motivation. Not wanting to be left behind in knowledge by the enlisted ranks, officer knowledge picked up as well.

With mission requirements resolved, next came the subordinate command requirement to identify training objectives and develop a written plan for achievement. From that came annual training plans, quarterly calendars, and weekly training schedules developed three weeks out. My visits to subordinate units not only included reviewing these documents, but also being shown requests for qualification ranges, ammunition, and all other logistical support.

Another document heavily enforced by Crowell and me was the Army Training and Evaluation Program (ARTEP). We discovered junior officers, and most enlisted soldiers had not been briefed on their respective ARTEPs. The ARTEP laid out the task, condition, and standard for every critical task a unit must perform to succeed in its field mission. It should have been the anchor point for identifying training needs, evaluating the success or failure of the training, and establishing the focus of future training. It was impossible for units to develop the annual training plan and implement the after-action review process effectively without the aggressive use of the ARTEP. Failure to include all officers, NCOs, and junior enlisted personnel in the existence of this document guaranteed failure to achieve full mission success.

These enhancements were implemented within weeks of Crowell's and my arrival in the command. Training programs soon reflected the missions, ARTEPs, and division mission statement. Achievement of goals and objectives, especially when set high, takes time. For that, subordinate commands were allowed a few months to bring themselves up to the division's newly established standards, provided steady progress was proven during our visits.

Crowell's core leadership team was rounded out by two brigadier deputy commanding generals (DCGs). The senior brigadier focused primarily on operations and training. The second was responsible for logistics and material readiness. Like me, the brigadiers preferred to go forward to the subordinate commands. Their logic was good. Unlike being at headquarters, with the subordinate commands, they were free of distractions and interruptions and could give total attention to the needs of the units visited. Seldom did they travel together as that would overwhelm the commands they were trying to support. It would also leave Crowell without the immediate availability of a deputy.

As a team, the DCGs and I worked on and solved most problems. Getting the commander's verbal or written approval for what we presented was never an issue. The strength behind what we were doing was that it always came from a team without being concerned about who had what rank. Our job was to support the commander, who was ultimately responsible for the mission, success, and future of the division in both peacetime and combat. There was always a standing invitation from the DCGs for me to join them when either of them visited subordinate units. Unless something prohibited my travel, the invitation was always accepted. The visits started with business with brigade commanders and their staff. Typically, after the initial meetings were completed, the DCGs would spend time talking to officers and I would break off to be with NCOs.

The pleasure portion for both the brigadiers and me was taking the opportunity to spend time with the troops, especially in informal settings such as in a day room or dining facility. Motor pools were also one of my favorite meeting locations. The reaction we experienced was always rewarding. It was different for the soldiers down to the squad level to see a general officer and the command sergeant major coming out together and working closely to make sure commands were getting the training they needed. The soldiers liked it. Often, we would be told that they did not know who were the DCGs and command sergeant major.

One soldier said he appreciated knowing whom to contact if a problem came up in the unit when the people between him and me could not fix it. This comment prompted me to let all brigade and battalion command sergeants major know my stance about soldiers having to bring problems to my attention because they could not get a resolution any other way. In both group and one-on-one meetings with subordinate command sergeants major, my position was made clear, "I will listen to them and will help if necessary. I do not ever expect to be told you would not listen to one of your soldiers. If that happens you will be the second problem I will be solving. I expect you to take the same stand with your first sergeants and they with their platoon sergeants." It only took a couple of "close encounters of an unpleasant kind" for all NCOs to start ensuring soldier needs were being addressed.

The coordination between Crowell, both DCGs, the G–4, and me brought resolution within the command to a widespread annual funding problem. Across the Army, senior commands were provided funding to conduct their missions. One of two problems will occur if funds are not properly spent as the year progresses. The less frequent of the two is funding running out early. The more prevalent occurs near the end of the fiscal calendar on September 30th when a stash of money exists that could have been wisely used throughout the year for necessities. If not spent, the funding is returned to the Army. The following year's budget is cut to the amount of what was spent. To prevent this, commands often found last-minute ways to spend the money, no matter how wasteful.

To ensure the funding was properly spent throughout the year, the division core leadership team and G–4 balanced the funding available with training missions, enhancements to training programs, facilities, and all other command responsibilities. All of us worked closely with brigade commanders and command sergeants major to identify critical things early in the year that required attention. We closed out the year exactly within the budget and with no wasteful spending to balance out the books.

We never had enough money for ammunition. This included individual weapons, machine guns, mortars, tanks, and artillery. Rifle marksmanship of sixty rounds a year and nine more for annual zeroing of the weapon, at best, only allowed for maintaining familiarization. Dry fire training only goes so far in improving basic marksmanship skills. It takes bullets into the targets to turn an unqualified soldier or marksman into a sharpshooter or expert.

Being in Germany afforded me good opportunities to train. The maximum advantage had not been taken so we started executing training development from the top down. Once programs were in place, we implemented training from the bottom up. At the company level, followed by battalion and then brigade, we devised effective training programs, not just splashy messes that consumed time, energy, resources, and soldier patience. Just as team capabilities are built on the proficiency of individual soldiers, battalion success was built upon the effectiveness of companies. No organization can be better than the combination of its components. The near-simultaneous assignments of several new brigade commanders turned into a plus. New commanders were easier to motivate than ones who had become complacent in their positions. To this day the maxim "no command can be better than its commander" remains an accurate statement.

Command readiness and proficiency within the division varied. All commands were "ready" for combat, some just more ready than others. That should never happen and all should be ready at the most proficient state possible. From various angles, all commands contribute to mission success. They have differing responsibilities, equipment, and technologies.

They also wear the same uniform, are subject to the same regulations, are required to maintain professional standards on and off duty and are expected to perform their duties to their utmost capabilities. With the exception of one group within the 3d ID, the NCOs were not pushing their responsibilities.

Like the various allied corps stationed throughout Germany, each command all the way down to company must be comfortable that the command on its left and right will not just hold ground but take the fight to the enemy. That goes back to 3d ID's "Rock of the Marne" legacy earned in World War I when the division held firm at the Marne River, blocking a major German offensive against Paris. Crowell and I continually made the point that every command and soldier was expected to live up to that legacy. In terms of Maslow's Hierarchy of Needs, that legacy and its effectiveness provided soldiers with self-esteem. We were determined to advance that self-esteem to self-actualization, all the way from individual soldier to division level.

The group that was the positive exception was division artillery NCOs. Working closely with equally good officers, they were the most aggressive and focused on doing their mission correctly. All senior officers and NCOs had served in Vietnam. Those years of being on forward observer teams and launching rounds as part of the firing batteries had served them well. They were still doing their jobs with precision. The attention they put on equipment readiness and facility maintenance was beyond reproach. If there was a shortfall within their operations, they only had to be told of it once.

My assignment as 3d ID command sergeant major came with mixed reactions among subordinate NCOs. Overall, the NCO corps was still broken and a far cry from the days of enlisted leaders like Sergeant Franco.

The negative impact of both Vietnam and the "All Modern Volunteer Army" was still being felt. Many NCOs, and officers for that matter, had learned to get by doing as little as possible, making as few decisions as necessary, receiving nice evaluations, and getting promoted on schedule.

Although I had a strong airborne background, I enjoyed opportunities to be assigned to brigades and divisions that were not airborne. The majority of command sergeants major within the division respected and appreciated me for moving up quickly because it empowered them. Backed by General Crowell, we set out to make further improvements to the NCOs' ranks. Quality brigade and battalion command sergeants major who were already in place were retained. Weak ones or those who were retired on active duty (ROAD) had a grace period to improve. If they failed, because of either capabilities or lack of willingness, they were replaced with soldiers who could and would do the job. Those who wished to remain "retired" got their wish, just not on active duty. The goal was to have quality NCOs leading the troops in all battalions and brigades. I knew there were people who did not like me, but I did not care. I did not like them either.

One of the enhancements Crowell and I implemented was the passing of information along from the company up to the division. Issues concerning training and care of soldiers were to be passed through command sergeants major. Even if the issues had been resolved at battalion or brigade level within specific commands, the problems might still be lingering within other commands. Identifying proven solutions was just as important as identifying problems.

Directly related to the achievement of goals and objectives for any command was the professional development of leaders. Within the commands, daily responsibility was on the commanders and senior NCOs. For the division, formal education of soldiers ranking from specialist to sergeant first class was the responsibility of the NCO Academy. After all the years of working with NCO academies, now, as the division's top soldier, I had one working for me.

My first visit to the division's NCO Academy was within days after my arrival. When the visit was over, the school cadre probably thought of it more as an unannounced inspection. I was not interested in slide briefings and a review of statistics. This visit was an examination of the facilities, training materials, lesson plans, and instructor qualifications. The facilities and training needed upgrading. Money was available and even designated at the division level but had not been requested by the academy. Most of the blame for this could have gone to the academy leadership, but the division staff was not completely off the hook. Training and logistics staff members could have left their offices to visit the academy and identify needs and ways to effectively spend the available funds. Their NCOs who attended the courses should have addressed the shortfalls to their supervisors and it should not have taken the new command sergeant major to get involved.

To the credit of the academy, the course instruction requirements and lesson plans being used were exactly as published by TRADOC. The problem was, like with all accreditation mandates, TRADOC had established minimum standards. Across the entire Army, those needed to be the same. However, TRADOC allowed the academies to adapt blocks of instruction and focus on the missions of the specific divisions. 3d ID was different from the 1st AD. There must be a specialization of courses and instructional intent. As proven when the 24th ID NCO Academy was developed, TRADOC had never been opposed to receiving, reviewing, and approving mission-specific training for any of the academies. My on-site assessments revealed the 3d ID NCO Academy had not taken the necessary steps to adapt their training problems to the commander's mission statement. That came to a quick change.

Then came the instructor review. Assignment to any training program, whether officer or enlisted-focused, should neither be a random assignment nor one to get a substandard soldier out of a line unit's way. Only the best should be assigned to develop subordinate soldiers. During my tenure in the division, the "best" not only meant knowledge and ability to present their subjects, but personal discipline as well. Any inability to work with soldiers and violations of the UCMJ on or off duty resulted in accountability and removal from the NCO Academy. The mandates I placed on the academy staff also came with accountability for failure.

Within a matter of weeks, the quality of NCO training moved up several notches. Upon graduation from the various NCO courses, the NCOs returned to their units with a thorough understanding of their responsibilities, expectations being placed upon them, the mission of their command, and where they fit into that mission. By no coincidence, the speed of upgrading the NCO Academy was parallel with commands improving their training programs. The two upgrades were interrelated. As with the academy, we worked hard on training needs throughout the division. Comments from both within and outside the command were made that the 3d ID had become a training unit. We did not take offense to this statement, because it was true.

In one way or another, the division was always involved in REFORGER. The majority of the time, the 3d ID was the good guys in forward-positioned locations. Our mission was to hold back opposition forces until reinforcements arrived. In this role, the 3d ID was often transferred from VII Corps to V Corps. This was part of exercising the reality that fluid situations created in combat may require a division to be assigned to another corps for command and control. A benefit to 3d ID's attachment to the V Corps during REFORGER helped solidify the unity of American commands in Europe. It was good for V Corps staff to understand the capabilities of and maintain a close working relationship with the division and corps on their southern flank. The same concept applied to us knowing the command to our north. For non-REFORGER operations, we remained with the VII Corps, especially when it involved infantry, armor, and artillery live-fire exercises at Grafenwoehr.

The 3d ID was the first to receive the new Abrams tanks. It replaced the M60 tank, which had proven itself in Vietnam and by the Israelis in the Yom Kippur War. Years later, while still being phased out of the US inventory, the M60 proved itself against Soviet-made Iraqi tanks in Operation DESERT STORM. The American and West German militaries had previously tried a joint endeavor to develop a single tank version to counter the then-perceived threat, Soviet T–62s. The effort fell apart, and the Germans moved on to replace their Leopard 1s with 2s. The Germans were focused on protection through speed and agility. Both versions of the Leopard fulfilled this intent very well.

Army Chief of Staff General Abrams was directly involved in the development of America's new tank. Not just because he was the best tank commander of his time, but also through his involvement in making sure development was done right, Abrams earned the honor of having the tank named after him. He had mandated expectations and tank criteria before his death from cancer while still serving as Chief. Among the technological advances, the Abrams tank came with advanced composite armor, nuclear, biological, and chemical protection for the crew, a multifuel turbine engine, and a computerized fire control system. Although weighing in at sixty-eight short tons, a limit specifically set by Abrams, the tank was successfully designed to maneuver quickly and engage targets without stopping. All the technology and advancements on the Abrams allowed it to surpass the Leopard 2 as being the best tank in the world.

General Abrams mandated a blowout compartment for ammunition storage. As a World War II tank commander, he knew well the consequences of tankers being trapped with ammunition when hit by enemy fire. This is a key vulnerability the Soviets never overcame even with their over-hyped T–72 tanks that was coming out at the same time as the Abrams and Leopard 2s. To this day, ammunition for the T–72 is stored directly below the operating crew. In turn, the British produced anti-tank weapons designed to cause T–72 ammunition to explode while blowing the turret off the tank.

When the Abrams tanks arrived in-country, I went with the receiving party to expedite delivery, work through any issues, and take possession of the equipment. Once our NCOs and maintenance staff confirmed everything was correct, we put our signatures on the receipts. Like every other piece of machinery in the Army, the Abrams tanks came with all the necessary manuals for training and maintenance. Our mission was to learn to use our internal and external resources together to understand how to use the tanks.

Our tankers were able to quickly adapt their M60 skills over to the Abrams. A few instructors and technical representatives came with the tanks. Fortunately, we had already sent several of our own officers and NCOs back to the States for operator and maintenance training. Our S–3s and assistant S–3s worked with the soldiers who had received the training to develop an education program. Next came integration with the rest of the division. At the same time the tankers were being trained in how to use the Abrams, the infantry had to train in how to keep up. Of the three infantry brigades, only one was mechanized, with three subordinate battalions. The mechanized and non-mechanized commands had already mastered coordinating with each other. We also had the M113s that allowed flexibility in deploying the infantry troops. Having all the M113s in one brigade worked well for training and maintenance. In the field, the brigade commander was always willing to attach his troops to another brigade if the mission required it.

On weekends the command sergeant major of the first receiving battalion and I went through the same training as the troops did during the week. Rather than feeling imposed upon that their time off was being spent on two relic NCOs, the trainers took pride that their bosses cared enough to spend their time learning as well. The battalion command sergeant major had an advantage over me, as he had been a tanker throughout his entire career.

Communication with the West German military was always very productive. Combined training turned out well. A major difference between American and German militaries is the use of doctrine. "Trying to determine American course of action by applying US Army doctrine is counterproductive. Americans don't study it themselves" was a comment initially attributed to a World War II German officer. There was merit to the assessment. Germans are organized and regimented. This is not necessarily bad as it generally works for them. Americans are much more fluid. War is chaos and the best battle plans tend to come apart very quickly after the fighting begins. The opposition also has a vote on how the battle will unfold. As proven under Truscott's leadership of the division in Sicily, American soldiers were at their best when the combat environment was the most chaotic.

Unlike the US military, West Germans were still drafting young men. Percentagewise the German draft population was much greater than the Americans before Vietnam, but the term of service was not as long. For the most part, it was indoctrination and skills training. With the Warsaw Pact threat at its eastern border, the West German government wanted to have as many military-trained men as possible to call into service should an invasion occur. There was a secondary benefit to what the German government was doing. It was enhancing its entry-level workforce by providing young men with discipline, focus, and competitive employment skills.

Exchanging officers and NCOs with the West Germans during field exercises was common. From this, we learned a lot about each other's tactical strength and application of forces. From each other, we learned a lot of lessons worthy of implementing in our own forces, including those at the company command level. Except for infantry, back then the easiest way to find an American command at night was to listen for the sounds of a generator running. The Germans dug a pit, using the ground as baffling, to cut down on the distance in which the generator could be heard. When West German tactics were better than our own, we implemented their procedures.

The Germans were much more focused on vehicle and equipment maintenance. This started at the private level and worked its way up through the chain of command. While our command level maintenance operations consisted primarily of soldiers rotating in and out on a three-year basis, the West Germans employed a lot of civilians who could be working in the same maintenance shops for decades.

As she had done at Garlstedt, upon our arrival at Wurzburg, my wife Margaret immediately began working on base with the soldiers' wives' support group. This quickly led to her involvement in the local West German community's ladies' auxiliary club. Margaret's ability to win people's hearts and confidence was impressive. More important, she never violated their trust. She was our ambassador, both on and off base. As part of our responsibilities, we have division, brigade, and community social events to attend. For me, accompanying Margaret was the best part of attending any event.

On base, Margaret was also an important set of eyes and ears. The nice family quarters we had been assigned allowed her to host gatherings and talk to people in a comfortable environment when necessary. Soldiers typically said nothing when a problem existed or, when asked, they would respond that everything was fine. Instead, they presented their frustrations to their spouses. Sometimes soldiers were the problem, including spousal or child abuse, alcoholism, or other misconduct. There came a point when the spouses realized the problem could not be ignored and they needed someone to listen. That's when they would come to Margaret. In the evening, she would report it to me. It was interesting how fast NCOs could move when an aggressive division command sergeant major knows more than they do about what is happening within their ranks.

Two community-related problems she brought to my attention involved official travel that participants had to pay for out of their own pockets. The first concerns spouses having to travel to take care of family support missions. The second concerned baseball and football teams representing their commands. Both groups were just looking for lodging when necessary and money for fuel. They were not asking for per diem, just compensation for their expenses. No one in the division was holding anything back, they just did not know about the problem. What I learned from Margaret was passed on to the G–4 logistics office. Once made aware, the G–4 officer worked with both groups, and compensation was provided.

One day, the community ladies' club took a trip to East Germany. This was not like the American military and family members crossing over the border under Soviet authority and sponsorship at Berlin's Checkpoint Charlie. This one was totally using civilian passports at a West/East

German crossing point. Unaware of it until it was over, I told Margaret she was lucky not to have been captured. They did have an amazing time.

Being in Southern Germany made it convenient for the family to travel to France and Italy. For the most part, all the major landmarks and attractions were in the cities. For visiting people, in both France and Italy, the countryside was the best place to go. When we left Germany, we felt like we had accomplished a lot, made a difference, and shared great experiences as a family. My next assignment was the desert of West Texas and New Mexico.

CHAPTER 14
Sergeants Major Academy

Since our time together in the 1st Ranger Battalion, Glen Morrell had gone on to serve as command sergeant major of the Recruiting Command, followed by FORSCOM. Now he was Sergeant Major of the Army and had a problem he wanted fixed. It was no secret to any of us that the Sergeants Major Academy at Fort Bliss had failed to adapt to the changing times.

Morrell was aware of my success after having been pulled from his battalion to build the 24th ID NCO Academy. We were still together at Fort Stewart, and our families, especially our wives, Karen and Margaret, were close. As Sergeant Major of the Army, he had personally seen the enhancements achieved at the 3d ID's academy. My next mission was to rebuild and reform what was supposed to be the Army's premier NCO academy.

Since the days of Lieutenant George Patton Jr. camping out on General Pershing's porch in his successful effort to be included in the Punitive Expedition against Pancho Villa in 1916, Fort Bliss had advanced to become one of the most modern bases in the Army. On the three sides is El Paso, Texas. The majority of Fort Bliss' land mass is in New Mexico.

Consisting of 1,700 square miles of real estate, Fort Bliss was the Army's second largest base. Only White Sands to its immediate north takes up more space. Combining the training areas of both installations, they might produce enough trees to complete one square mile of Bavarian countryside. The Gates family was in for a major change of scenery.

In 1972, the Sergeants Major Academy was established on the north side of the main post. Buildings left were left over from World War II, and the Korean War were still there. In those buildings, the academy had been designated. The only neighbors we had were Army reservists on one side of us in the summer months and year-round aviation operations at Biggs Army Airfield on the other. This area itself was ideal as it allowed us the opportunity to be isolated from the activities and traffic associated with Fort Bliss. This was good for training, but routine maintenance at the compound had been neglected and the academy facilities needed a lot of work. Families of cadre had pretty good quarters, but they were outdated, including my own.

For administrative oversight, we had a colonel assigned to the academy who was still getting settled in when I arrived. His role was to provide top cover and exercise UCMJ authority whenever necessary. He made it clear up front that I was really the commander. He was not being lazy or derelict in his duties. He fully supported the intent that the academy was to be operated by NCOs, for NCOs. The colonel and I immediately found the right combination of an officer and senior NCO relationship. He was my champion for engaging Fort Bliss leadership in getting the logistical and maintenance support we needed. Only once did he ever have to administer disciplinary action. A student had alcohol in the barracks, which was strictly forbidden. When it came time for his next assignment rotation, he credited us with providing the knowledge on how to maximize the use of NCOs.

Shortly after my arrival at the academy, the TRADOC command sergeant major came for a visit. For some reason that I never figured out, he made a comment about training black soldiers. I responded, "Soldiers are soldiers, no matter what color. If you're in the US Army, you're a soldier and that's how people should be treated." He said that I was "talking about equal opportunity." I replied, "That is exactly right. I cannot see color because I do not look for it."

Our colonel's engagement with the base's commanding general and my engagement with the command sergeant major always produced results. We went out of our way to work with them, and they returned the courtesy. Every time Morrell came to Fort Bliss, I extended an invite for the command sergeant major to join us. Courtesy quickly developed into a professional relationship. The support we needed started coming in fast. For every class, the base's commanding general accepted our invitations to speak with the students. Upon graduation, he emphasized their importance in becoming battalion and brigade top soldiers. They loved it. For many, this was the first time they were able to have an open forum with a two-star general. On his first visit, the general saw the condition of our dining facility. He immediately dispatched base engineers to refurbish the building.

The best answer the colonel and I could figure out as to why previous support from the base was lacking was because base leadership had not been asked to participate in academy operations. The commanding general had funding that he was willing to dedicate to our needs. He was honored to have the Army's top NCO academy on his base. That positive attitude radiated its way through the installation. Even the post-exchange supervisor was willing to assist. Repeating what happened with the 24th ID NCO Academy, a simple visit to the supervisor resulted in a trailer being permanently parked at the academy for students and cadre to buy hygiene and other small items. With support coming in from the entire base, logistical and facility maintenance issues did not take long to resolve. It's amazing how much support can be obtained by just bringing needs to the attention of people who can help.

The biggest problem was the quality of training. Today, the Sergeants Major Academy's declared mission is to "Provide the Army with agile, adaptive senior enlisted leaders of character, competence, and commitment to be effective leaders. Those leaders, grounded in Army and Joint doctrine, exploit opportunities by leveraging and applying Army resources." The Academy declares its vision as "the premier professional military education (PME) institution focused on providing the Army with agile and adaptive senior enlisted leaders." This mission and this vision are exactly what Morrell was determined to achieve. The NCO Corps was not as broken as it had been at the end of the Vietnam War, but it still had a lot of work ahead. Developing the best sergeants major possible and sending them back out into the field was the best way to get the job done.

A two-way unwritten contract existed between students and any school they attended. On the student's part, the expectation was for maximum effort to apply themselves in learning. On the school's part, it was to provide an environment where the students can learn and to provide an education that will enhance their future chances of success. If a comparison were to be made between today's mission and vision statements against what the academy was providing before Morrell's intervention, there would be serious deficits.

The curriculum and lesson plans were flawed. I attended the academy eight years prior and assessed then that the training was out of date. Nothing had changed. The same lesson plans were being used, and the academy was still teaching Vietnam era philosophies and solutions. The instruction was dry. The cadre and students knew it. This was an eight month "check the box" operation to qualify NCOs for rank advancement. About nine hundred senior NCOs attended the academy the year I was there. A typical course would have about 180 students.

Most of the students came for the training and worked hard while there. The problem was they were working hard on studying what was about fifty percent junk. What skills they had upon arrival at the school were pretty much the ones they had upon graduation. The academy was adding little to their professional development. It did not make sense to keep teaching the wrong things, knowing they were wrong.

The curriculum and quality of the lesson plans were only part of the problem. The cadre needed upgrading. About half of them were not top-notch trainers. A good selection process did not exist. For the most part, the academy was getting what Army personnel sent. Once at the academy, they went straight into the classroom. There was no "train the trainer" program in place to prepare them or to determine who was not cut out to be an instructor. To train recruits, drill sergeants must be certified through a very rigorous course. Yet to train the Army's most senior enlisted soldiers, no preparation was completed. The academy's belief was that if they were senior NCOs, then they must know how to conduct training. Some of the instructors did not have a clear idea of what they should be doing.

We also had the issue that many instructors had been at the academy for so long that they lost perception of what was really happening in the ranks. Most students who attended the academy had more knowledge about the current Army than those homesteading instructors. That was embarrassing. Students from the field having more knowledge than their instructors in a specific course was never a unique issue, but it should not have been such a predominant one.

A command in combat that is performing poorly can be pulled back to the rear, assigned new leaders, and retrained. General Bradley did that for the 90th ID in World War II. He later assessed that the 90th arrived in the European Theater as one of the worst commands, but through training and effective leadership, it became one of the best. Bradley's option of pulling the unit off line was not available to us. We had three classes already on the deck, each about two months apart, with more already scheduled to come in. We had to fix our problems while engaged in the mission.

Army Personnel Command was stopped from sending anyone to the academy until after I conducted a personal interview and made an official acceptance. My network of contacts was exercised to ensure we were not picking up someone else's problems. I knew all the division command sergeants major. Their knowledge and checks with subordinate brigade NCOs resulted in my receiving accurate verbal evaluations, not those over-inflated written evaluations being placed in personnel files.

Instructors already assigned, and those coming in, were required to attend a "train-the-trainer" course. We also established behavior standards expected of every member of the cadre. This included in the classroom, on duty, and off duty. During the first few months, the colonel's authority was exercised many times to approve outbound transfers and host several retirement ceremonies. There was no loss in the departure of these people as they had been occupying positions ready to be filled by soldiers who wanted to work. With Glen's backing, we had no problem receiving a competent replacement cadre.

Along with fixing the cadre problem, we set out to develop training that was relevant to the real Army. Most of the curriculum and lesson plans came from TRADOC but needed to be adapted, upgraded, and updated. Just because a lesson plan was TRADOC approved ten years ago and still on file did not make it relevant to present times. Conducting a team analysis, we identified three basic categories for the existing courses: maintain as currently developed, maintain with enhancements, or completely trash. In addition to the experience of the cadre, we used another immediately available resource. All students were either master sergeants or sergeant majors. Just like with that Ranger soldier out in the California desert who came up with the idea of calling the fire department, we solicited, listened to, and acted upon their recommendations. The most critical group of students were the ones nearing training completion. They had gone through most of the courses.

Simultaneously by identifying courses for modification or elimination, we identified potential courses for future training. Recommendations went through a review process. We were not going to replace old junk with new junk. Even before we completed finalizing the revised eight-month training program, we started developing lesson plans and identifying needed resources for courses we knew would be included. We were developing advanced training based on what was currently happening in the field. We were not applying yesterday's knowledge. This was not done in a vacuum. We reached out to NCO schools throughout the Army and other military branches.

Once we put all this together, we had to send the entire package to TRADOC for approval. We did not hit them blindly as we kept TRADOC staff informed as we progressed. Here we ran into a problem I did not see coming. Unlike when I was in the 24th and 3d Infantry Divisions, the engagement of TRADOC staff for approvals was not producing results. When Morrell came down for each graduation, often that same TRADOC command sergeant major would accompany him. As a target of opportunity, I'd address the lack of written approvals. Even then, we did not see results. The inaction of his command had had nothing to do with my response to his ridiculous racial statement made after my arrival. It was a leadership problem, starting with the command sergeant major.

Something else was starting to happen within TRADOC. Rather than committing themselves to approving what we were developing, TRADOC's recently assigned staff members found staying inside their personal safety zones more comfortable. What we got was an answer by no answer. They were either protecting themselves from something potentially going wrong in the future or just too incompetent to do their jobs. From my observations and the senior NCOs attending the academy, I realized this was a growing problem throughout the Army. I did not know it time, but I would be involved in the fix years later.

Our solution at the Sergeants Major Academy to the TRADOC approval problem was to make copies of the documentation we sent forward, along with the periodic resubmissions. Meanwhile, students were provided with the blocks of instruction we, and they for that matter, had determined as necessary for their professional development. Continuing to build upon cadre and student input allowed us to make the training even more robust and current.

In addition to enhancing the NCO Corps through the academy, Glen had another mission for us. He wanted to know the pulse of the Army. Having all these senior NCOs away from the daily demands of their commands afforded him an opportunity. The academy always had writing and presentation assignments. Instead of the typical reports that we would receive from every class, Glen wanted us to have the students write papers addressing what they saw as the state of the Army. The papers were developed both as individual and team projects. From the papers, briefings were developed. The students knew up front that the best-developed reports and presentations were going to be forwarded to the Sergeant Major of the Army. Not only were they students in his academy, but they were also his think tank on how to make the Army better. The students went all out to make it work. Several good things came from this effort. The SMA received a lot of outstanding papers, which helped him in his job. The students appreciated being heard all the way to the Pentagon and knowing perhaps their input would make a difference. They also received a reinforcement reminder that requesting input from subordinates can help in both decision-making and boosting command morale.

For outdoor training, we buddied each student up with one who was more familiar with the subject. Weapons training and qualifications were a prime example. Our goal was not to produce expert shooters but to educate leaders who would be supervising the trainers in the battalions and brigades. To do that, we took the students back to the basics of shooting and worked forward. We could do dry fire, weapons assembly, and weapons malfunction training on the academy grounds.

The academy was so remote at the time that we could also do M16 and .45–caliber qualifications behind the building. For familiarization with more powerful weapons such as the M60 machine gun and the M203 grenade launcher, we went to McGregor Range, a few miles to the north. As senior NCOs they were already expected to be proficient with M16 rifles. That expectation did not always prove true as many of the students were not combat arms branch. Many of the administratively oriented combat service support troops only held a weapon long enough to complete annual qualifications. In these situations, as in all training, we put students with strong capabilities with weaker ones. It was the Ranger buddy system all over again. Continuously, the case was that students who needed assistance in certain areas proved to be ideal coaches in other training blocks.

There was another historically neglected area that required our attention. Students could take their spouses to Fort Bliss while going through the academy but as isolated as we were there was not a lot for them to do while the students were in class. In the early 1980s the upscale businesses, family restaurants, and blocks of nice homes that now populate the area north of Fort Bliss were not there. For the most part, there were a couple of strip malls, gas stations, fast food restaurants, pawn shops, beat-up apartments, bars, motels that received much of their business from the bars, and a lot of desert. Being the closest to the academy, this was the most geographically convenient area where the students with spouses found lodging. It was in this environment that spouses would spend the day waiting for their other halves to finish training.

Margaret was upset that the spouses were being ignored. Always a person who saw a way to turn a negative into a positive, she organized the academy spouses of cadre and students into a unified group called the ULTIMA Ladies Association. The way the wives operated was impressive. They did not beg for anything, just politely asked for assistance. They requested a place where they could meet and do their own training. As we had no shortage of old buildings, the colonel and I agreed to let them have one. The base commander came through with engineers and facility maintenance to make the building a nice place.

On their own, they developed several training programs. The spouses were married to senior NCOs and remembered the days of living in run-down apartments and trailers while their husbands were at work, on field exercises, or on long-term deployments. They knew what other spouses were now going through. They talked and shared ideas, and collectively, they possessed massive experience, knowing how to work with organizations like the Red Cross and Army Emergency Relief. The spouse group received a lot of training in public engagement and learned how to say the right words to the right people at the right time. One of their group's favorite activities was going to the hospitals to visit patients.

Their training included taking care of children. In military families, children often do not receive adequate attention during all the transfers and deployments while attending new schools and developing new friendships with strangers who are going through the same experiences. Because our spouses had dealt with these issues daily, they were the subject matter experts on working though the problems. Like everything else they were addressing, the spouses developed and refined their own lesson plans for long-term use after they had moved on.

Having been the wife of a division command sergeant, Margaret knew how to engage in senior leadership and work through bureaucracy to get help. She felt that once she learned about a problem, she had a responsibility to get engaged in addressing it. Many of these wives would go on to be first ladies of brigades and divisions. Margaret was coaching them for the future. She also built a network of contacts that was pretty much intact during my future assignment as Sergeant Major of the Army.

Aside from their own classes, the spouses reached out to us for support. Some of our classes, like operations security, more commonly referred to as OPSEC, were ideal for a joint session. They cared about their family's security, and they understood the dangers of talking in public about their spouses' upcoming deployments. We also warned them about reporters coming to family homes when the service members were away, especially if something major was unfolding.

The spouses group also did some traveling. One of their requests concerned travel assistance for going to Juarez, Mexico for inexpensive shopping and lunch. We got buses to transport them and added some cadre to the trip for security. The wives stayed together, looked out for each other, and returned safely. Fortunately, this was before the days of the powerful crime cartels that now exist there.

When Sergeant Major of the Army Glen Morrell came for the graduation ceremonies, his wife Karen was almost always with him. Her delightful personality made everyone feel special. Because of the travel time between the Pentagon and Fort Bliss, Glen and Karen would make each visit a two-day event. This provided Morrell time to talk to the classes and review the staff enhancements being made to the program. These two-day visits also allowed Margaret and I to spend time with two very close friends.

Karen would accompany the wives' group during the day's activities and talk to spouses both individually and as a group. One of the talks was about uniting with other spouses during times of tragedy. Karen and Margaret knew firsthand about having husbands deployed on combat tours. Their husbands returned home alive and with all their limbs. Both wives had watched friends who were not so fortunate. Spouses and families will receive notifications of deaths while a battle is still going on.

The effectiveness of the wives' group had a very positive impact on the students and cadre. Each day, when the students entered the classrooms, they could focus on training, knowing their accompanying spouses were fully engaged in an environment of mutual respect and meaningful activities. The wives were preparing themselves for their future responsibilities and enjoying themselves while doing it. The ladies had developed a total winning situation for themselves, their spouses, and the academy.

Times have changed. The spouses who accompanied the students in 1984 were all ladies. Today, both the sexes of students and of spouses are going to be mixed. That's progress. Hopefully, progress included what our spouses accomplished, and adjusted to the changing times. A lot of things they accomplished were worthy of being carried forward.

Morrell and I were determined not only to upgrade the quality of the academy but also the structural environment. Those World War II buildings gave way to a state-of-the-art learning center that exists today. Getting into modern facilities was important, but that was secondary to creating an academy that could adapt to a changing world. The academy was released from being trapped in a time that had passed and developed into a flexible educational institution.

CHAPTER 15
US Forces Korea Command Sergeant Major

The Sergeants Major Academy was on track with meaningful instruction. Students felt they were not just punching tickets for advancement. More spouses were accompanying the students as word about the wives' club had filtered back to friends awaiting attendance at the academy. After finishing their eight months of training, the graduates and their spouses were prepared to meet future demands.

Glen Morrell's initiative to challenge the students to become a voice for reform was working. In every cycle, the best presentations and reports were forwarded to the Army Chief of Staff as realistic proposals for his consideration. The depth of research and the organizational development that went into the reports were impressive. Subsequent classes were provided with the reports of their predecessors as professional development templates to build upon and as standards for them to not only equal, but surpass.

On the home front, my family was enjoying Fort Bliss. After a year in the American Southwest, what had seemed like barren land when we first arrived from Southern Germany was proving to have its own special beauty. "Oh, my God" were the words I uttered when the unexpected orders came in assigning me as command sergeant major of US Forces—Korea. Margaret's initial reaction was the same. There was no prior indication that I was even being considered for the position.

My first concern was uprooting the family once again, which would be the fourth time in five years. Margaret quickly started looking forward to the experience; she thought seeing the Republic of Korea (ROK) would be interesting. At first, I thought she was trying to make it easier on me, but I then realized that she was really looking forward to living in Asia. Considering my first two tours of duty in Asia, it was obvious Korea would be a significant improvement. My family arrived in Inchon, South Korea, by commercial airline. A security team was waiting at the terminal and took us to our new home. Our personal belongings and car arrived a few weeks later. We received briefings, and despite the country being run by a dictatorship, we realized that Korea was no longer the country portrayed on the television series *M.A.S.H.*

My new boss, General William J. Livsey, served simultaneously as commander of the US Eighth Army, US Forces–Korea (USFK), ROK/US Combined Forces Command, and U.N. Command. Years later, the Eighth Army would separate from the command, and a lieutenant general would become the commander. On his watch, General Livsey had all of these responsibilities, and as his command sergeant major, my responsibilities included them as well. This tour of duty was the first for me in several ways. All previous assignments had been working within the US Army, with some experience with other American service branches. Prior work with Vietnamese and German military forces was pretty much limited to their armies. Now I was working with all four service branches of the American and South Korean militaries.

It did not take me long to realize Livsey's decision to bring me on without an interview was based on several factors. My name was known as he had been the VII Corps commander in Germany while I was working as the command sergeant major of the 3d ID. He also had a network of general officers to talk to, including all my old bosses. Glen Morrell's opinion was also valued. The usual progression for a command sergeant major having finished time at a division is to advance to a command led by a three-star general, then to a four-star command. For me, in South Korea, both were going to happen, simultaneously. Livsey was as tough as they came while being very much into the training and care of the soldier. He had served two aggressive combat tours in Vietnam. His other command assignments included 8th ID, the US Army Infantry School, and the Third Army. He was my kind of officer.

Our combined mission headquarters was Yongsan in the heart of Seoul. It consisted of two compounds separated by an east-west municipal boulevard. A footbridge connected the bases, and we recognized the need for a vehicle bridge that was eventually constructed. Most military operations were in the north compound. Eighth Army headquarters occupied a massive three-story building built by the Japanese for the same purpose. The empty spaces above the third floors where Americans had chiseled away the giant Rising Sun emblems over the entranceways still showed when I was there.

Across the street from the Eighth Army headquarters and beyond the parade field was the USFK/Combined Forces Command–Korea (CFC)/United Nations Command Headquarters. It was a modern two-story rectangular administration building with no unique characteristics other than the painted color, earning its title as "The White House." Previously, all command sergeant major duties had been performed out of the Eighth Army building; Livsey wanted me to have an office closer to his primary office. As I intended to be at the headquarters only ten percent of the time, my recommendation was to place a general-purpose tent on the parade field outside his office. The life span of that proposal was zero minutes.

The mutual respect of the American services toward each other in the USFK was outstanding. There was no rivalry as to which one was the better service. We all knew we had a mission to do and depended on each other's branch to fulfill that mission and survive should combat operations occur. The same respect existed between the Americans and South Koreans. We learned a lot of policies and techniques on how to work with forces other than the United States.

The Seventh Air Force was located at Osan, thirty miles to the south. The US Air Force (USAF) lieutenant general assigned to the command also served as USAF Korea Deputy Commander and conducted most of his administrative command operations from "The White House." K–2 was another major Air Force base outside Taegu. That base also included South Korean fighter jets with at least one section always on alert.

Most Marines were stationed on the east coast at Camp Mujik outside Pohang. My visits to Camp Mujik always went very well. Their sergeant major had an excellent program ensuring the troops were training how they would fight. I only had to tell Marine NCOs to do something once. They might come back with a better way to get the mission done or questions in search of clarification, which I expected of them and any other subordinate. They never tried to provide an excuse for not getting the job done. The Marines worked closely with their ROK counterparts. Joint missions were accomplished with seamless integration. At a distance, the ROK Marine emblem could be mistaken for our own.

US Navy forces in Korea operated the southern port at Busan. At Yongsan, we had a Navy one-star admiral and a limited number of sailors assigned to the USFK staff. The Navy carrier fleet stayed at sea most of the time, close enough to provide support in time of hostilities. Seeing how their service members lived compared to Army soldiers in the field brought back memories of the song "The Navy gets the gravy, the Army gets the beans."

The American and South Korean navies worked closely with each other. In times of war, the US Navy would take the lead in the battle. Years later, South Korean frigate patrols proved to be more effective than US forces in dealing with Somalian pirates operating off the coast of Africa. Americans would board the pirate mother ships, capture the crews, and go through the trial and confinement process. The South Korean Navy, having tracked the pirate assault boats back to their motherships, simply blew everything out of the water.

Like the US Coast Guard, South Korea's Maritime Police Agency does a remarkable job. The Maritime Police Force was always armed and ready for a sustained gunfight. Headquartered west of Seoul at Incheon, their subordinate commands patrol the entire South Korean coast looking for North Korean infiltrators. The top of their mission priorities was protection against sea-borne saboteurs of the country's many ports. Members of the Maritime Police Force were good at their jobs, and they knew it. The North Koreans knew it too, which was why so little sea-borne sabotage had occurred.

Although South Korean forces were led, equipped, and financed by their government, as part of combined forces in time of conflict, they were under the control of Livsey. Our declared mission was to stop the North Koreans from coming south. The reality was that we were also keeping the South from going north. If the two Koreas went to war and China, Russia, the United States, and every other nation of the world stayed out of the fight, South Korea would have dominated the air, land, and sea battle sectors within a week. North Korean military leaders knew this as well. On the north side of the Demilitarized Zone (DMZ), North Korea had positioned long-range artillery weapons and Soviet-procured missiles dedicated to attacking the Greater Seoul Metropolitan Area (GSMA). Kim Sung-il knew allied counter-artillery and air forces could wipe out his offensive sites. By numbering those sites well into the thousands, by the time the last one was destroyed, the GSMA would be reduced to rubble, and several million residents would be murdered. Citizens of the GSMA were de facto hostages. They were allowed to live and prosper but would immediately be executed in time of open conflict.

North Korea's artillery strategy also ensured their forces could never invade down the peninsula's western corridor. South Korea's economic miracle had resulted in GSMA skyscrapers in every available piece of real estate between the mountains and coastal wetlands. Neither wheeled nor tracked vehicles were ever going to get through all the rubble that would be created by North Korean artillery. Meanwhile, stalled the allied war machine would have annihilated North Korean forces.

Much of my time was spent fulfilling my Eighth Army responsibilities on the front side of my Korean tour than on the back. The Eighth Army had units spread throughout the country. The two largest were the 2d ID and 19th Theater Support Command (TSC). The 2d ID and its subordinate command were located north of Seoul, with the largest concentration at Camp Casey, directly east of Tongducheon. Between Tongducheon and Seoul was the merging metropolis of Uijeongbu. Our artillery battalions were spread out at various camps. The readiness mission of the Eighth Army is to work with South Korean forces to blunt a North Korean advance, while receiving and leading incoming US Army commands. The 19th TSC, redesignated the 19th Expeditionary Sustainment Command in 2005, was located far to the south at Daegu. Headquarters in Camp Henry, the 19th TSC was a combat support command on steroids.

The Eighth Army NCO Academy cadre had no reason to be surprised when I walked through their door. The same problems existed here as in the 3d ID and the Sergeants Major Academy. The classrooms and dormitories were run down. Instructors and students were not wearing the proper uniforms. The lack of educational quality meant this was only a check-the-box operation. Students returning from the academy to their units should have been holding their heads higher, been more professional in their appearance, and been smarter. There was a significant difference between what it was and what it could have achieved. For me, seeing a repetition of the same problems in Europe, the US, and now Korea has become a test of patience and a source of disgust.

The academy commandant and I had a calibration meeting on the instructor and student self-discipline issues. To his credit, he understood my instructions and went to work making adjustments. In our initial meeting, he showed me the documentation validating his efforts in submitting requisitions to the Eighth Army staff to obtain funds to improve the facility and receive the necessary equipment. For him, the headquarters was a black hole. Whatever he sent in disappeared, and nothing ever came out. I discovered that Eighth Army G–4 staff was part of that government-wide problem of tight-fisting the money during the year to ensure the budget did not run out prematurely. Like everywhere else, they approved everything and anything near the end of the fiscal year to get the money spent. Working with Livsey, the money was freed up for the academy and all other requisitions that had been conscientiously submitted. In the end, we had nice training areas and dormitory rooms.

We got slide projectors and training aids. There were no weapons assigned to the academy. Requiring students to bring firearms from their units violated numerous principles of weapons accountability. The academy already had a certified arms room, so with the commanding general's support, the academy received its inventory of M16 rifles. In support of the commanders, where the students came from, their assigned weapons stayed in the unit arms rooms where they could remain properly accounted for and secured. Expecting students to bring automatic weapons from all corners of the peninsula and individually transport them back when classes were completed because the academy failed to have its own inventory defied logic.

Some of the old instructors had to go, but most were good once they received retraining. People changed their mindset, and things improved. There was a little complaining from the students that they were no longer allowed to go downtown to the bars, so I reminded them that they were there to work and learn, not to have fun and be impeded by hangovers. They decided that was right, and party time could wait until they returned to their units.

The typical classroom was built for thirty to forty students. Time has proven that twenty to twenty-five works the best. That became the standard. Broken down into progressive phases, about 150 students could attend the academy at a time. They enjoyed going there and developed professional relationships with fellow NCOs. We allocated ten slots per cycle for Korean NCOs, with three conditions. They had to be volunteers, their commands had to be approved, and they had to be fluent in English. Having ROK NCOs in the Eighth Army NCO Academy was a great educational asset. As always before, my goal was to tailor academic training to the environment in which we were located. Among these students, we had subject matter experts sharing knowledge of their country, people, geography, and military capabilities. American graduates of the NCO academy returned to their units with facts, not assumptions, about South Korea.

In one training cycle, the academy had upped its standards so much higher that I could tone down my aggressiveness. That did not stop me from attending classes whenever I felt the urge. Soon, my unexpected visits produced different results. Instead of staff and instructors becoming concerned with my presence, they were happy introducing me to the students whenever I walked into the classrooms. They also did not hesitate to address training and equipment needs that would allow them to perform their duties better.

During my subsequent visits to their respective commands, the graduates of our improved program were proud to talk about their positive experiences. There was a trickle-down effect because better NCOs were returning to their units. First sergeants, sergeants major, and commanders told me about the positive improvements they witnessed in returning soldiers. Later as Sergeant Major of the Army, I went back to the Eighth Army Academy, and it was rewarding to see the standards we had established were being maintained.

In the CFC, I had a counterpart Korean sergeant major who worked closely with me. It started out a little blurry, but after a couple of months, we became good professional friends. I would give him information about things he needed to look at and get fixed that I had seen in my own observations. He had a sound system for making corrections. A pet project of ours was developing a ROK NCO Academy. Reports had come up from the ranks of the ROK about NCOs who attended the Eighth Army Academy and how much they learned. My Korean counterpart asked for help and support in developing a similar academy for his leaders. The Eighth Army Academy helped in every way possible. It took longer for the ROK academy to meet our professional standards. They had to start their climb from the bottom of the hill, take a role of leadership in their changing environment, adapt our curriculum and lesson plans, and develop their own. The important thing was, like everything the Korean minds put focus on achieving, they did not stop until success was achieved. In the end, the two academies mirrored each other.

My initial visit to the Eighth Army NCO Academy was almost immediately upon my assumption of duty in Korea. Getting out to the subordinate commands took a couple of days longer. First was the 2d ID. The division commander was a good leader and emphasized the need to "Fight Now." He understood this was going to be a come-as-you-are event with no preparation time. All the required equipment was within the subordinate commands.

Division, brigades, and battalion leaders were informed they could expect my visits to really be inspections of their command training, maintenance, and care of the soldier programs. I also informed the command sergeants major to expect me to walk into their offices anytime. Throughout the division, yearly training plans and calendars had been properly completed. Quarterly requirements were being achieved, but with one problem. Many units had fudged the three-week training schedule publication requirement down to two weeks, and some were failing to publish the schedules a week in advance. There was no excuse that this had not been corrected by the battalion commanders, command sergeants major, and S–3 operations staffs.

The soldiers who were paying the price for this failure were the unit-level leaders assigned to conduct training. Development of training schedules three weeks out alerts the assigned instructors of their upcoming missions. In turn, this allowed time for research, lesson plan refinement, requisition of training aids, and equipment coordination. For almost all training, lesson plans already exist, but need to be adjusted to fit the instructor, students, and mission. While preparing to conduct the class, the instructor still had daily job requirements to complete. Just the knowledge that I would be conducting unexpected visits and examining the training fixed the problem. Soon, all commands had their schedules posted on the bulletin boards for three weeks out.

The division's Aviation Brigade at Camp Stanley included Pathfinders who would jump first to establish the drop and landing zones for incoming paratroopers and helicopters. A special connection happened as soon as I made my first visit when they realized my background. During our open discussion time, a soldier revealed to me that they had not been receiving jump pay because the brigade was not affording them the opportunity and aircraft to perform regular jumps. Their senior had been trying to get it done, but the aviation unit had closed them out.

Before leaving the unit area, I spoke to the command sergeant major. He had been focused on helicopters and his aviation troops, apparently thinking jump pay for the Pathfinders was not in his department. I made it his department, informing him, "Your pilots, your flight crews, they get flight pay. So why don't the Pathfinders jump once a month and get jump pay? Those soldiers belong to you." I emphasized he was not taking care of his soldiers, and if I found out about this problem on my first visit, the question was how many other problems existed that this NCO was failing to resolve.

Upon returning to "The White House," I reported the situation to Livsey, and he agreed that the pay issue was the first problem. The fact that those soldiers were not being taken up into the air to maintain their unique skills, as well as confidence in themselves and their equipment, was the second problem. From Eighth Army staff, both the G–1 and the G–3 colonels were ordered to get involved and fix the problem. Prior to being summoned into Livsey's office, they did not know the problem existed.

Next came the 2d ID commander, who subsequently engaged his own command sergeant major and the aviation brigade commander. It is never pleasant when a four-star general knows more about what is happening or not happening in a colonel-level command than the people responsible for leading it. Ultimately, all the attention that came with this revelation ended up on top of the command sergeant major who did not think taking care of the Pathfinders was his responsibility.

The very next day, the Pathfinders had jump pay orders and jumped. We always had the money. They got paid in less than two weeks and received some back pay. This had been a long-term problem that was fixed in one day because someone listened. A short while later, at the beginning of the day, my secretary told me I had a visitor. She did not know who he was, except he was a soldier. That was good enough. I replied, "Let him in." He was one of the Pathfinders and had come over to express his appreciation that I had stood up for them. I told him that when a soldier has a problem, it is his NCO's problem to fix it or run it up the flagpole. He said he would never forget that. It is amazing how people will come back and talk to you.

Margaret had a similar experience when she went to the orphanage that housed children born to Korean women, fathered by American service members. The Koreans did not want anything to do with them. USFK had established a good-sized secure compound with a living area and a school. USFK was also funding the orphanage and paying the wages of the facility manager. Margaret went to visit because she was always concerned about children's education and their preparation for the future. That place was a mess. The building was decrepit, and children of various ages had no beds to sleep on; if lucky, they had a sheet to place between themselves and the floor. The commodes could not be properly flushed, and the showers did not work. They lived in filth and their diet was mainly rice with very little fish or meat. The children were born into a world in which they had done no wrong. They were suffering the consequences of no one caring for them. Margaret got so upset that she called me at my office in Yongsan. I could not go immediately, so we both went the next day. It was terrible. I went to Livsey and said, "Sir, you have got to look at that place." He told his aide to block his calendar, and we immediately went there together.

On the spot, Livsey fired the manager, and a Korean-American was subsequently placed in overall charge. Once again, proper funding of the facility was always available, just not being distributed and used correctly. That changed on the spot. The G–4 was assigned to get real beds, mattresses, and bedding for the children. The engineer commander was assigned to fix the building. Once they began the mission and personally witnessed what the children had been enduring, the engineers were not satisfied just to work duty hours to make things right, and they worked late into the evenings and on weekends. Additional volunteers came from every direction, including the wives' club. Within a month, the orphanage was completely redone. Quality education materials and instruction were made available to the children. Toys were brought in. Caring staff were recruited to operate the facility and work with the children. We started an awareness program back in the States about the children being available for adoption and had a lot of success.

The looks on the children's faces were filled with happiness when they realized what was being put together before their eyes. Three real meals a day and their own beds were the two things that had the greatest impact on them. As for Margaret, every time she went back to visit the orphanage the children would come running and give her hugs.

Separating North and South Korea was the DMZ stretching one kilometer into each country. At the border in 2d ID's section was Panmunjom, location of the original peace talks. It was also where Livsey went when he needed to conduct United Nations business with the North Koreans. This was also the compound where two American officers were murdered with axes by North Korean soldiers while supervising the trimming of a tree that was impeding observation just eleven years prior.

Periodic tours of Panmunjom were provided by the Eighth Army's Morale, Welfare, and Recreation Office. Margaret went on one of those tours. Whenever tours were held, visitors were escorted into the buildings that were used for negotiations. As part of the tour, one of the guides walked around to the North Korean side of the table to speak to the audience. Repeating her earlier trip into East Germany, Margaret followed him. Fortunately, while these tours were ongoing, the North Korean guards remained outside the closed doors. This was not a problem, as it was also part of the tour. However, if she had opened one of those doors on the buildings north side, she would have been taken prisoner. It would have been embarrassing for General Livsey to have to have special negotiations with the North Koreans just to get Margaret back.

A few months into my time in Korea, the M1 Abrams tanks came into the country. Some of the armored soldiers were familiar with the Abrams from their stateside assignments. For me, this was an opportunity to play team coach. When the tanks arrived in Korea, there was a surprise among some of the staff that this infantryman knew how to operate them. Those weekends spent learning armor maneuvering in Germany had paid off. Working with the senior NCOs of the armored brigade and the division staff, we were able to identify and walk through all the necessary steps for training and logistical support. My knowledge was still fresh as it had only been two years since the 3d ID had received the Abrams. The difference was that in the 3d ID, we had to send our requests for support up through the chain of command with the expectation of receiving timely assistance.

As team coach, I was tracking all support requests that were being sent to Eighth Army staff and everyone involved knew it.

Then came the international event that would change South Korea forever, the 1988 Summer Olympics. For the first time in twelve years Soviet, Chinese, and American athletes were able to compete with each other. The 1980 Summer Olympics in Moscow had been boycotted by the United States and West Germany due to the Soviet Union's invasion of Afghanistan. In turn, the 1984 Summer Olympics in Los Angeles had been boycotted by the Soviet Union, East Germany, and several other Warsaw Pact nations. North Korea and a few of its allies boycotted the Seoul Olympics. Because Soviet and Chinese representatives were participating, we had a better feeling of comfort that the ever-unpredictable North Korea was not going to try to sabotage the events. Even still, security was a priority, and some American security companies were contracted to assist.

In emergency response mode, Livsey maintained increased vigilance among all the joint US/ROK forces. The USFK and ROK emergency response centers remained operational and in continual contact with each other. South Korea's image and future meant that the Olympics had to be seen as a successful civilian event. As it turned out, civilian law enforcement and security had everything under control and did a remarkable job. Being on the ready was not wasted effort on the part of the US and ROK military forces. It's always better to be prepared for something that does not happen than to be caught flatfooted. The unofficial winner of the 1988 Summer Olympics, based on medals won, was the Soviet Union. The irony is that this was the Soviet Union's and East Germany's last time to participate. By the time the 1992 Olympics came around, both countries had ceased to exist.

For Americans stationed in country, there was no question as to why we were there and what service members before us had contributed, during and since the Korean War. This fact was not lost on the Koreans, as it had been in other countries where Americans had fought and died to rescue them from tyranny. I witnessed this not only as a soldier but as a father. When my family left the base to go out into the community, Korean people treated us very well, even though they did not speak English and we only knew a few polite words of Korean. Melissa did not accompany us to Korea as she was now in college. In the summer months, she would come to visit, and the three Gates ladies would shop in Seoul together. My understanding with Margaret was that she would only drive our car on base, so I pretended not to know she was driving it into the city.

Margaret was right in her belief that being stationed in South Korea was going to be a good experience for our family. We were enjoying ourselves in a beautiful country, witnessing history in the making, believing we were making a positive difference, when orders came through for my next transfer.

CHAPTER 16
Sergeant Major of the Army

My interview with the Army Chief of Staff (Chief) General Carl E. Vuono for my selection as Sergeant Major of the Army went well. We shared a lot of common denominators in both our backgrounds and thoughts about discipline, soldier development, training, and unit readiness. Vuono graduated from West Point in 1957. As a field artillery officer, his three combat tours in Vietnam were with the 1st Infantry, 1st Cavalry, and 82d Airborne Divisions. Putting steel on target was a standard he continued to exercise in his non-combat assignments. He was heavily focused on training. General Vuono also pushed for the fundamentals to be followed, and every change that was approved had to be put into the system immediately and enforced. Over the years, he had served several assignments in TRADOC, where he served as commander prior to becoming Chief of Staff.

Following the interview, on the return flight to Korea, I had no thoughts one way or the other as to what would happen with the selection. A lot of projects were still ongoing. The selection announcement came with a one-month lead time before the requirement to report to the Pentagon. Being the ultimate professional, Vuono kept the decision on a close hold until he told Livsey, who then informed me. My responsibility was to tell

Margaret and Lauren. Margaret's ability to adapt to sudden environmental and mission changes was always amazing.

The next four weeks were filled with transitioning responsibilities to my interim successor, visiting units, and fine-tuning NCO training programs, as well as dealing with household goods and auto shipments, dinners, and social calls. Program managers knew they were not off the hook for maintaining the standards we had achieved. Each one of them knew that, in my new position, I would be conducting follow-up visits. They also knew that as a combat veteran, once I secured a piece of real estate, I expected it to be held and maintained.

During the month of transition from Korea, I randomly jotted down notes of objectives to accomplish as Sergeant Major of the Army. On the plane flight to America, with my Korean responsibilities completed, those notes were worked into a comprehensive plan of action based upon items I identified as needing attention through the years. The number one thing I wanted to accomplish was getting a good program for the promotion of soldiers to noncommissioned officers. The program needed to be refined, and the right soldiers needed to be in the correct positions. They needed to earn it, not just have longevity in the Army. I learned good ways to accomplish this on my way up the ranks.

Another reason for working on this list was to be ready for my first duty meeting with Vuono. Through my interview meeting, Vuono's reputation and personality, plus preparatory guidance received from Morrell, I had a good idea of what the Chief expected from his sergeant major. I knew Morrell had already been working on many of the same objectives, especially in NCO development and unit training. We were going to be like two relay runners passing the baton on a sprint.

Any professional having an initial meeting with their new boss should always walk in with a list of goals and objectives. A good boss is going to have a list as well. Differences in the lists should not be viewed as conflicts, but rather opportunities for discussion and mutual expansion. Overlaps in the lists are immediate opportunities to create synergy for the common good. Ultimately, the boss's list will prevail if different objectives are not mutually accepted. This concept should occur all the way through the ranks, starting at the squad leader.

My family spent a couple of days setting up our new home in Fort Myer in Virginia. Being in Washington worked out well for our family. The city itself is a museum of American history. Within the city there are hundreds of museums and places to visit. Margaret and Lauren had opportunities to buy clothes in D.C. that they couldn't find elsewhere. Margaret was a sharp lady and knew how to dress according to the function or ceremony she was attending. The established protocol determined what level of dress uniform I was to wear. This was fortunate because I did not know style other than for casual wear and farm work.

The initial meeting with Vuono lasted for two hours and answered the question as to why he wanted me as top soldier. At the forefront, Vuono was determined that Morrell's accomplishments would not be lost because of a successor who was only interested in a prestigious retirement tour position. Then came the issue of NCO leadership. Vuono had done his homework, and it became obvious he had conversations with my former bosses, starting with Armstrong. He stated his appreciation of my style of leadership from the battalion on up and wanted it pushed throughout the army.

For the Chief to specifically identify the importance of battalion leadership, confirmed I was going to be working with a senior officer who had not lost contact with reality. Brigade and higher commands must be focused on battle-space operations and maneuvering of forces. The battalion commander and command sergeant major have the greatest impact and create the environment that will determine the success of their subordinates. The combat successes or failures of the brigades and divisions reveal how effectively the battalions trained their troops, developed them into effective combat teams, and enforced equipment maintenance.

As for our two lists of objectives, we had differences but no conflicts. During the meeting, the Chief's list and mine merged with both authors' appreciation of the other's insights. We agreed that most of my time would be spent away from the Pentagon. The big issues he wanted me to focus on were training and discipline. He wanted training information to get down to the soldiers at the squad and platoon levels to improve their performance. Discipline was also taken down to the company level. The Chief specifically emphasized haircuts and uniform appearance. The Army's top general usually doesn't talk about these things, but the problem was serious enough that he made it his (and my) problem.

My role was to find out where the communication breaks were and force the NCOs to pull the information through the blockages and get it distributed. Vuono explained that during unit visits, I was not only examining compliance with his directions, but also, I was examining compliance with the directions of army, corps, and division commanders. My presence with the units was not only to be visits, but also assessments. Vuono was creating an environment where officers and NCOs from the battalion up would obtain and implement senior command guidance. He was taking the expression "NCOs are the backbone of the Army" to a new level.

One Army problem was that commanding generals would say how they wanted things done, but their instructions were never delivered to junior officers and mid-level NCOs. New policies and directions need to be understood and implemented instead of just being words that end up being ignored. Vouno wanted to get the information disseminated to all levels, not just with the updated training manual but with all guidance and regulations. That is where the Chief wanted me to come in. I was going out with the sergeants major, first sergeants, platoon sergeants, squad leaders, and individual soldiers.

Our first meeting closed on an interesting note. Vuono told me that when it was him and me behind closed doors, we could call each other by first names. My comeback was that I would use both of his first names, "General" and "Sir." We then embarked on a great four-year run.

We immediately updated the Army field manual for training, FM 25–100, *Training the Force*. Even though the previous version was only two years old, the emphasis Vuono and I placed on training warranted publication of new guidance. Leaving no room for misinterpretation, Vuono's instructions were included in the opening statements: "The Army training mission is to prepare Soldiers, leaders, and units to deploy, fight, and win in combat at any intensity level, anywhere, anytime."

I did not believe the Army Pentagon staff had ever experienced a Chief of Staff and Sergeant Major of the Army so involved in developing any field manual. The entire document of the completed manual was only sixty-six pages. We had a support staff to bring everything together and develop the graphics. With accuracy, brevity, and clarity, we made our standards known. Anyone who knew us and knew the way we talked and wrote recognized our work throughout the manual. The document came out in 1988, and it covered everything soldiers, leaders, and units at all levels needed to accomplish for the execution of effective training. The final words inside the back cover capped the manual well: "Training, then and now, must be the Army's top peacetime priority—it is the cornerstone of readiness." FM 25–100 was the standard we demanded of everyone in the Army.

During the introductory week of organizing my office and meeting the various Pentagon organizations, I addressed the first discipline problem. It concerned Vuono's own Pentagon staff. A sizeable number of officers and NCOs wore tennis shoes with their uniforms in the mornings and evenings. My inquiry as to why this was happening resulted in being told that the hard surfaces of the subway platforms, parking lots, and Pentagon corridors were tough on the pads of their feet. To prevent undue pain, they left their dress shoes at their desks and wore cushioned sportswear while coming to and leaving work. This answer created a flashback to the days in the Vietnamese swamps when leather boots rotted off our feet. Twenty years later, we have rear-echelon staffers avoiding the pain of wearing dress shoes on hard surfaces.

Immediately upon briefing Vuono about this issue, his executive officer was called into the office and ordered to type out orders for the Chief's signature, bringing the practice to an end. The NCOs complied without comment. The only complaints were from officers who had been at the Pentagon for several years. I wished we could have brought in a child's plastic swimming pool, filled it with swamp water and leeches, and had these whining staff officers stand in it for several hours to comprehend the real meaning of foot pain.

In the same week, I came across another blatant violation of soldiering standards. As small as Fort Myer and co-located Henderson Hall were compared to combat arms posts, I found a way to get in a ten-mile early morning run. While running, I observed a formation of soldiers conducting PT while doing what was supposed to be push-ups. There were many variations, no cadence synchronization, and several people standing behind the formation talking to each other. I stopped my run and walked up to two people in conversation standing behind the gaggle, who turned out to be a lieutenant colonel and a sergeant major. I asked them what they thought they were doing and what this thing was that slightly resembled a PT formation. This was the US Army Band, which made me hope they did not play their instruments as uncoordinatedly as their exercises. The lieutenant colonel arrogantly asked, "Who are you?" The sergeant major stated, "He is Sergeant Major of the Army Gates." At least one had looked at their command bulletin board since the beginning of the week. Both then received from me on the spot training concerning the proper way to conduct organized PT, and I ordered them to participate as well. Beginning my run again, I heard the lieutenant colonel say that he was in trouble.

Shortly after arriving in my office, I received a phone call from an NCO wanting to know why there was concern. I advised him that those in charge were not supposed to be standing back, running their mouths. They must ensure that exercises are done correctly and participate in the training. Someone else grabbed his phone and said, "Sergeant Major, Sergeant Major, we will get it fixed and do all those things correctly." He was the operations sergeant major and told me his name. At about 10 a.m., they came to my office with an order mandating PT in compliance with the field manual and that everyone present would be participating in the exercises. To the credit of the command, they were doing everything right every time I ran past their PT formation in the future.

Like the Pathfinder soldier who came to my office in Korea, one day, a soldier from the band saw me on Fort Myer and asked to speak with me. He told me how much he appreciated someone forcing his supervisors to include themselves in what became professional PT formations.

PT formations were not the only thing that needed to be cleaned up on the Myer/Henderson bases. I was not specifically looking for deficiencies during the morning runs or evening walks. A mental note was taken when I observed a problem, and the organization responsible would be contacted. The whole complex started changing for the better with lawns being mowed and wind-blown trash no longer accumulating in the corners. A lot of people thought this was great and apologized that I had to waste time checking on them instead of being where they knew I wanted to be—out with my soldiers. I did not see it that way. They, too, were either soldiers or civilian employees serving the Army mission. My responsibility was to ensure they lived and worked in a sanitary and professional environment. Addressing the cleanup did not take me away from anything, other than the time it took to make a few phone calls.

The first major change in the office layout was placing metal canteen cups next to the coffee pot. This was to remind everyone that we were soldiers first, and our mission as Pentagon staff was to always anchor on the lives and needs of ground troops. If someone wanted coffee in my office, a metal canteen cup was what they had to use. They could not bring in their own ceramic or paper cup. Vuono liked the idea. A ground soldier himself was at home using a canteen cup. His staff had mixed feelings. Some took to it. Others did not like it and would turn around and leave. This worked for my benefit. If they came only for a free cup of coffee, their presence was a waste of my staff's time. In time, those with a reason for coming into the office adjusted and used the canteen cups.

Each week, unless he was on travel, Vuono would have a three-star staff meeting, consisting of G–1, 2, 3/5/7, and 4. The Vice Chief of Staff of the Army (Vice), who was also present, would lead the meetings when the Chief was unavailable. If not on travel, I also attended the meetings, sitting 45 degrees from the Chief and across from the Vice.

One visitor who occasionally walked into my office unannounced was Chairman of the Joint Chiefs of Staff General Colin L. Powell. I could tell how long the visit would be by how much coffee he had poured into the canteen cup on his way in. He and I were good friends even though he was the boss of bosses. We shared two things: humble upbringing and aggressive combat experience. In Vietnam, Powell had proven himself well, never losing focus on his mission even when the helicopter he was in was shot down. He had a philosophy that all people in positions of responsibility should remember: "You are no longer a leader when subordinates stop feeling they can bring to you their concerns."

His visits were usually focused in two directions. Sometimes, he looked for unvarnished ground truths about what was happening in the field. Other times, he came in to relax and chat. The Chairman just wanted to talk to another ground soldier in a world of dealing with politicians and polished bureaucrats. Sometimes I would get a call to come to his office for the conversations. Vuono never had a problem with these meetings as both men respected and trusted each other. The Chief appreciated that Powell knew where to reach out for sanity calibrations in the often-dysfunctional environment of the Washington beltway.

Periodically, Vuono and I had private meetings where we would have mutual sanity checks and vent our frustrations. In one of these meetings, I vented my disgust about the Pentagon long-termers. Having easy access to their assignment officers and NCOs, they were choosing to stay away from the combat divisions. There were many officers and NCOs doing this "homesteading" while doing as little work as possible. It was from this group that most of the staffers wearing sports shoes had originated, and caught my original attention. Everything I subsequently observed did not improve their image. They had created for themselves an environment inside the Pentagon reminiscent of antiquated weapons systems on display in various museums around the country—they did not work and could not be fired.

Vuono agreed with me and informed the G–1 that he wanted these people to rotate into the real Army. Most of those with over twenty years of service retired rather than accept transfer. Many of them had already lined up civilian jobs and were staying on active duty only to increase their retirement pay. That group was already ROAD and prevented other soldiers who wanted to work from getting an assignment where they could make a positive difference. Within months, we had gotten rid of the homesteaders and started witnessing a massive improvement. The hard workers appreciated the departure of their lazy peers and especially their deadweight seniors. Good officers and NCOs screened and selected from outside the beltway started coming onto the Pentagon staff. Overall morale took a sharp rise. This afforded me the chance to spend time bragging on the staff and telling them about the good jobs they were doing. None of this was artificial because they were doing a great job. Knowing the Sergeant Major of the Army noticed their work, took time to thank them, and bragged about them to the Chief, which caused them to work even harder and improve their performance even more. In the military, we call it the result of professional leadership.

Ridding ourselves of the dead weight produced a secondary effect of encouraging discussion among the various staff. Instead of people avoiding integration with other staff, we had an increase of energetic people anxious to learn and make a difference. It did not take long to see that our "who else needs to know" philosophy had positive results. Stovepiping of information never accomplished anything.

Vuono and Powell were not the only senior officers with which I had a close relationship. Future Chief of Staff Lieutenant General Gordon R.

Sullivan was a key member of our Pentagon team. He and I were very good friends before he was a brigadier. He always did a fantastic job and had a deep concern for soldiers. Having an outstanding combat record himself, his ultimate objective was always to work on what could improve the ability of soldiers to fight wars and come home safe. When the time came for him to receive his well-earned fourth star, he asked Vuono and me to pin the stars on him. Mine was his first salute in his new grade. I then told him he owed me $10 since he had given the person who first pinned/saluted him when he became a O–1 (second lieutenant) a dollar. Now that I saluted him as an O–10, he owed me ten dollars. A week later, he came into my office with a ten-dollar bill on a plaque. The engraved metal plate stated, "From GEN Gordon Sullivan to SMA Gates." That plaque is on my wall today.

My relationship with SecArmy John O. Marsh was not as close and sometimes he irritated me. To emphasize the importance of FM 25–100, *Training the Force,* Vuono had designated 1988 to be "The Year of Training" and it was coming to a close. Marsh was walking down the Pentagon hallways with the Chief and me when he asked what we should call the following year. I immediately said, "The Year of the NCO." Marsh responded, "Gates, you are always pushing the NCOs." He may not have meant it that way, but I took it as a compliment. After he moved on to his office, the Chief and I discussed the idea. Both of us agreed. "The Year of Training" should be followed up by recognizing the leaders who are primarily responsible for executing that training. We wanted to remind, challenge, and empower them. 1989 was officially designated the "Year of the NCO."

As in previous positions, I spent only about 10% of my time in my office. I was mostly out running around with the units, trying to get things done, get firsthand information for the Army, and to enforce standards. Even at the Pentagon, most of my time was dedicated to resolving issues discovered in the field. Patton had pointed out that it is always better for the senior to go forward, rather than have subordinates come to the rear. My going forward to the various offices showed respect for other people's workload, gave them the comfort of talking within their workspaces, which often included access to the files and reports they needed in our discussions, and provided time for my visiting with people. I put out the word not to treat myself like a dignitary, but as a sergeant major determined to support the Army's mission and the soldiers who get the job done. We got a lot accomplished, and those visits always resulted in conversations that benefited our common concerns.

Then came encouraging communications with counterparts in the other service branches. My time with the joint service command in USFK helped in this effort. If what we were doing pertained to the Air Force, Navy, and Marine Corps, then they needed to know. Before then, many our people would not tell the other services what they were doing. This was especially true with the Marines. I was the wrong person to discover this attitude. The two organizations that constantly worked together for the benefit of each other were the Army and the Marine Corps. We would fuss and harass each other a little bit. In combat, we worked closely together, and there was no reason it was not happening at the Pentagon. The inter-service enhancement that was subsequently created produced much input from the other forces, which really assisted in developing standards and strategies. We began seeing changes in our information assessments and intelligence reports. No one branch has exclusive knowledge.

As Sergeant Major of the Army, I worked closely with Army and Marine Corps NCOs. The whole group of us had good relationships. We were always coordinating with each other. We were in the early stages of computerized communications. These were great for sending detailed messages, but no electronic communications should ever be a substitute for direct verbal conversations. Face-to-face is the best, but impractical over long distances. Telephones work fine, as does video teleconferencing. Anyone who thinks e-mails and texts are the ultimate solutions to effective communication does not understand the basics of leadership.

The senior command sergeants major had already mastered a requirement that I previously forced on brigades and battalions when visiting their units. When a problem was presented, they were expected to have a proposed solution. There were several reasons for this. The most obvious was that no one wanted to hear simple complaining. If they had taken the proper amount of time to identify the consequences of a problem thoroughly, they also had time to develop proposed courses of corrective actions. This took us directly to the second reason. Developing proposed solutions requires working with the people ultimately affected by the problem. The byproduct of this requirement was subordinate leader development.

Credit must be given to the division level and higher command sergeants major I encountered during my Pentagon years. All of them were top of the line. Not only did they possess impressive credentials, but they, too, had been selected after rigorous vetting. Their commanding generals wanted the best and settled for nothing less. Discussion of problems and proposed solutions fit very well into our senior command sergeant major conference calls. Ideas could be shared, and if one command had found a solution to a similar situation, others could benefit from a peer's experience. It could also have been a problem within other commands that had not been exposed to the senior sergeant major level. They had time to do research and find out if something was being kept from them by subordinates. If the problem were widespread, we would look for the common denominator and identify solutions. We would look at, analyze, and approve anything new that came up as a team. The next step was my taking it to the Chief for approval, disapproval, or guidance on what he wanted included in the resolution actions. Until the Chief lifted the responsibility and passed it on to his G Staff, I owned all problems identified.

In keeping with this, Vuono would periodically call Pentagon action committees to address specific problems or develop a way for improvement. There is an inherent danger with committees. If not properly controlled, their counter-productivity is directly correlated to the number of people on them and the inflated egos of the members. The way around this was to identify a project lead responsible for setting the agenda, moderating the meetings, and documenting ongoing progress. Setting the agenda did not mean the project lead rammed self-desires down everyone else's throats. The project lead also needed to leave his or her inflated ego outside the conference room. Like everyone else within those action committees, my recommendations were no more valuable than the next person's. Everybody's ideas were put on the table. None were discounted until we finally came up with a solution.

It was from this process that Vuono implemented two changes that affected the entire Army for decades to come. First concerned utility uniform. The Battle Dress Uniform (BDU) that was being issued to soldiers at that time proved itself as a functional change from the olive drab fatigues that had been used since the late 1950s. Much thought and excellent design went into BDU development. The camouflage mixture of woodland colors, the large leg pockets, and the shirt draping over the trousers were instant hits. Those leg pockets were able to store all kinds of things. My preferred use was for NCOs and officers to carry their leader notebooks and for all soldiers to carry their skills qualification manuals. If idle time needed to be filled, NCOs could train from the readily available soldiers' manual.

The second change concerned field rations. Switching over from the canned C-rations to the newly developed Meals Ready to Eat (MREs) took time. As food, C-rations served the military well in World War II, Korea, and Vietnam. The plastic bagged MREs were easier to carry, especially when broken down from the original large pouch bags. The essential items could be packed well into rucksacks. However, the flavor of the food was boring at best. I proposed to spice it up with Tabasco sauce. The team toyed around with a potential flavor additive and came to the same conclusion I suggested at the beginning. In time, MREs did improve, and the tabasco sauce never went away.

Ever since the days of the first Sergeant Major of the Army William O. (Bill) Wooldridge, seeking the advice of the Army's top soldier has produced positive results. With Vietnam in full blaze, Wooldridge was shown the new concept of material to be prepositioned in anticipation of future needs. After examining the extensive list, he stated that something was forgotten. The G-4 representatives responded that the list had vehicle parts, soldier equipment, medical supplies, and everything the soldiers in combat could possibly need. Sergeant Major Wooldridge pointed out that they had forgotten to include ammunition.

We held an annual senior command sergeant major meeting at Fort Myer, lasting five to seven days each year. We would start with intelligence, operational readiness, logistics, and numerous other briefings to set the stage. Then we went into the work sessions to identify things that needed to be done to improve the Army. Refined ideas were returned to the Pentagon, not to be put in a records file, but to become action items. Progress was shared with the sergeants throughout the coming year. A formal status presentation was provided at the next annual meeting.

Like all my jobs in the Army, the best part was being with the troops. I spent ninety percent of the time away from the Pentagon and an information sergeant accompanied me on all trips. His job was to make sure I had everything necessary, to take down notes wherever we were, and to maintain a travel summary. The information sergeant sent out word that "dog and pony shows" were forbidden during my visits. Those things take too long to organize, especially because of all the ridiculous rehearsals. If I were going to visit a company to watch or learn about some specific operation, the division command sergeant major would want to see a rehearsal. Before him, the brigade and before him, the battalion. With all the slide development, rehearsals, briefing adjustments at every level, several people would have spent days preparing. If the event involved witnessing troops trying to work through a field mission or situation, of which I needed to learn of potential problems, they would have rehearsed it at least three times before my arrival, negating my effort to personally witness shortfalls that needed support.

Receiving the ground truth was the only way I could know what was going on in the unit. Before leaving the Pentagon, I reviewed the mission statements and Unit Status Reports to find out how the commands to be visited were doing. My goal was always to see if there was anything I could do to aid them. Those polished briefings and operational exercises never helped. On the briefing side, a one-page "issue, discussion, recommendation" paper was the best means of addressing a problem and easiest for the information sergeant to include in the trip report. A well-developed issue paper was also the best method of encouraging subsequent dialogue that would clarify what I needed to do to assist.

Like me, Vuono spent as much time as possible away from the Pentagon. This was possible because daily it is the Vice who runs the Army Pentagon staff. The Chief runs the Army, and he can do that from anywhere. I would accompany Vuono as much as possible.

On our joint trip to the Soviet Union, President Mikhail S. Gorbachev's reforms were taking place. The positive relationship between him and President Ronald W. Reagan continued into the subsequent Bush administration. We knew there were elements within both the Communist Party and Soviet military who opposed Gorbachev and wanted to go back to the old ways. Their personal empires were rooted in the oppression of the population and a less-than-friendly attitude toward America and Western Europe. The Warsaw Pact was very much in existence, even though everyone on both sides of the Berlin Wall knew the North Atlantic Treaty Organization (NATO) forces would reign supreme in an open conflict. Concerning the Soviets, Vuono and I viewed this visit as, "We'll meet as friends, but will also be ready to engage you on a field of battle."

Vuono and I were assigned separate escorts. The colonel assigned to me was a Komitet Gosudarstvennoy Bezopasnosti (KGB) intelligence officer. The professional relationship between Vuono and me had captured his curiosity. For that matter, the Soviets did not understand the American NCO Corps at all. Vuono and I discussed this before making the trip. We decided to have some fun and play with their minds. We knew we would be monitored, even by remote listening devices, while having private conversations. Rather than a concern, we saw it as an opportunity to shape their conclusions about American Army leadership. Vuono liked my plan to show no emotions other than a cold, aggressive attitude.

For the 11 p.m. changing of the guards, the Soviets hosted us at Lenin's tomb, where his body was preserved in an airtight glass-covered coffin. When Vuono asked me what I had thought of it, I said, "It was an outstanding ceremony. Their soldiers looked awesome, almost as good as my soldiers." Continuing to play our roles at the Soviet Army headquarters compound, Vuono asked me, "Did you ever think you would be inside these walls and see the guards?" I responded, "Not as a spectator. I always looked forward to coming in by parachute to kick butt." I spoke loudly enough to ensure my KGB officer heard my comments.

At the evening dinners, there was lots of vodka. My escort asked me, "Why do you not drink?" I responded, "As the sergeant major, I am always on duty. The general can drink. Even though he is in uniform, he is off duty. As his senior enlisted soldier, I must always be fully alert."

The Soviets took us up into a museum room where they displayed awards and decorations from militaries worldwide. Everything from all branches of the United States military was there. How they came to possess a Medal of Honor was beyond me. I asked that question once back at the Pentagon. We never got the answer. No one knew how the Soviets obtained all those awards and decorations.

The tour of their basic training camp was interesting. A lieutenant colonel was teaching recruits the skills of synchronized marching. My escort asked for my opinion. My response was, "You have senior officers doing the jobs of junior sergeants." Seeing a lieutenant colonel doing that training was amazing. Not since the American Revolution, when Friedrich Wilhelm von Steuben introduced drill and ceremonies to the Continental Army, has anyone other than NCOs been responsible for conducting training. Officers need to come around to check every now and then. But they have many other important things to do and teaching basic soldier skills is not one of them. This confirmed that in a firefight, taking out Soviet officers would leave their troops confused and ineffective. Vuono had a good time talking to Russian officers, but they would not let me speak to any NCOs.

Our trip to the Soviet Union was cut short. We received word from General Sullivan that we were needed back at the Pentagon immediately. For security reasons, we were not told why, but we had a good idea. Panamanian dictator General Manuel Noreiga had been stirring up trouble.

The Soviet Army Chief of Staff accompanied Vuono to our departing plane. I was standing at the bottom of the ramp. The two Chiefs were several meters away, and I saw the Soviet pointing his finger at me. Once we were airborne, Vuono told me his Soviet counterpart said about me, "That man is the meanest looking person I have ever seen. If all your sergeants are like him, then I'm glad we never went to war with each other."

Vuono then mentioned that he had seen me violate my pledge of not smiling while watching the Soviets working their T–72 tanks. It was a slight smile, but a smile, nonetheless. I explained that in trying to impress us with their latest and greatest in tank technology, the Soviets showed us their biggest maneuver vulnerability. T–72 crews had to stop and stabilize before engaging a target. In an unforced error, the Soviets had provided us with a key to defeating their state-of-the-art tank on the battlefield.

While on the move, our Abrams tanks could not only hit a stationary target, but also one that was moving as well. The German Leopard could do the same. NATO Forces could defeat the T–72s by forcing continual movement or denying the crews adequate time to stabilize and prepare for fire. This information was not only important to NATO armored troops, but also to those armed with anti-tank weapons. Vuono agreed, having exposed this T–72 vulnerability was something worth smiling about.

CHAPTER 17
Low and High Intensity Conflicts

American humorist Will Rogers observed that the United States is the only nation in the world that waits until it is in a war to prepare for it. President Ronald Reagan broke that trend. Oriented toward the collapsing of the Soviet Union, Reagan's financial investment in the military far exceeded the entire Warsaw Pact coalition. Reagan used his Strategic Defense Initiative, more commonly called "Star Wars," to consume much of Moscow's attention focus. Meanwhile, he advanced other investments in technology and equipment enhancements. The US Navy had over 600 ships, including a dozen aircraft carriers. The Army's Abrams tanks and Bradley fighting vehicles were the best in the world. Gorbachev's reforms in the Soviet Union made the world wonder what the United States was going to do with all its announced military advances. The years 1989, 1990, and 1991 answered those questions.

What completely flew under global radar, both physically and figuratively, was the development of F–1 Stealth bombers. Foreign spies would have been better off if they had not written off as crackpots people who were reporting unidentified flying objects over Nevada's top-secret Area 51 and Tonopah Test Range. Within those ultra-secret research and development sites, the US Air Force was conducting pilot training and fine tuning a war machine that was a generation ahead of anything else in the world.

At the Pentagon, we made sure those technological advances were properly fit into the three types of doctrinal forces tailored to combat missions: light, heavy, and special operations. Using a historical perspective, Reagan's 1983 Grenada "Operation URGENT FURY" invasion was a light force operation that required quick mobility into the combat zone and fast success once on the ground. Europe's 1944 "Operation OVERLORD" D-Day invasion was a heavy operation that required time for build-up and extensive use of tanks, artillery, multiple uses of air support, and massive naval support both in transport and bombardment.

Light forces, such as the 82d and the 101st Airborne Divisions, and Special Operations Forces, such as the Green Berets and Rangers, are most effective in heavy operations when senior commanders do not use them as standard ground troops. It is also critical for senior leaders to understand that there are times when Special Forces should not be used in lieu of heavy forces.

For all conceivable situations requiring American military forces, contingency plans for execution exist within the Pentagon. We do not wait for a crisis to develop and determine how to respond. The entire package had already been developed, including troop composition, strengths, logistical needs, transportation requirements, intelligence analysis, and everything else. Sometimes, when something unexpected occurs, an existing package for another operation can be quickly adjusted. This is what happened with URGENT FURY.

A well-developed plan was already in place concerning Panama. The Canal was too critical for US security and world economic stability to be ignored. Once a well-paid CIA informant during the Carter Administration, dictator General Manuel A. Noriega had been able to seize power in Panama in the early 1980s. Noriega was a prime example of the adage, "Power corrupts, and absolute power corrupts absolutely." He was using the entire country of Panama as an extension of his ever-growing criminal empire, including racketeering and illegal drug operations. All overland drug transportation from Colombia to North America had to go through Panama. Because of all the international shipping traversing the Panama Canal, Noriega was involved in drug trafficking throughout the globe, especially Western Europe.

When Panamanian opposition candidate Guillermo D. Endara won the presidential elections, Noriega voided the ballot process and declared himself the Supreme Ruler of Panama. President George H.W. Bush recognized the deteriorating situation and reinforced American military presence within the Canal Zone. Per the treaty, dating back to the days when President Theodore Roosevelt Jr.'s administration succeeded in the construction, the canal was under American control.

Noriega should have seen the writing on the wall when US Army General Maxwell R. Thurman was appointed chief of Southern Command (SOUTHCOM). "Mad Max" Thurman was tough and had long since proven himself as America's ultimate warrior of the 1990s. An artillery officer by trade, Thurman had multiple Vietnam combat tours, fathered the "Be All You Can Be" slogan when he was in charge of the US Army Recruiting Command, served as the Army's Vice Chief of Staff, and was commander of TRADOC when he applied for retirement. He had good reason for wanting this retirement, but when President Bush called on Thurman to lead SOUTHCOM, Thurman put his personal life aside and answered the call to duty.

Noriega numbered his own days by standing on a stage, waving a sword, and declaring war on America following his puppet General Assembly's December 15, 1989, declaration of war on America. This, and the subsequent killing of unarmed US Marine First Lieutenant Robert Paz by Noriega's military, triggered President Bush to give General Powell orders for invasion. Per Powell's determination, the invasion mission was designated Operation JUST CAUSE. It suited Powell to realize that even detractors would have to refer to the operation as "Just Cause." Then came the application and enhancement of the go-to-war standards established by Reagan's SecDef Casper Weinberger. Powell had served as an assistant to Weinberger and later directly served Reagan as National Security Advisor. Powell had retained the lessons learned from Weinberger and used them wisely when he became Chairman of the Joint Chiefs. He made sure all the Weinberger Doctrine points were covered as he advised President Bush and current SecDef Richard Cheney on the increasing certainty that US forces would need to go into Panama.

Within the standards of the Weinberger Doctrine, President Bush announced his four justifications for the invasion of Panama: safeguarding the 35,000 US citizens who lived and worked in Panama, ending Noriega's drug enterprise and thus protecting citizens of the US and Europe, enforcing the treaty between President Carter and President Torrijos that guaranteed America's use of the Panama Canal, and protecting democracy and human rights in Panama. Knowing that this was going to be a low-intensity operation against a far weaker and disorganized military force, we developed a limited operation scenario.

The best way to determine the effectiveness of an organization is how well it works in the absence of the Chief. Even though Vuono's and my return to the Pentagon from the Soviet Union had only taken a day, Army staff were well on their way to activating troops and resources, coordinating with our fellow service branches, and bringing everything together. Led by Vice Chief of Staff Sullivan, Pentagon staffers were at the top of their game. Witnessing how well they were working made Vuono and I glad we had cleared out the Pentagon dead weight two years earlier.

This was a team operation. My job was using the command sergeants major communications network to ensure Army troops were getting the support they needed and conducting on-site visits to ensure everything was coming together. In those five days between Noriega waving his sword and the US military answering the war declaration, I was impressed with the efficiency displayed by all service members, from privates to generals. All the joint training with the Air Force was paying off. The troops from Fort Bragg quickly moved themselves and their equipment over to co-located Pope Air Force Base. Load crews worked closely together, and the planes were made ready. All Army commands from units outside of Fort Bragg received the same professional care from the Air Force. Everything was made ready, knowing that at any point in time, President Bush had the option to order us to stand down. A distinguished combat veteran himself, there was no backing off in this Commander in Chief.

On December 20th, 1989, the American assault on Panama began. The close coordination between the US Army, Air Force, Marine Corps, and Navy commands was textbook. The XVIII Airborne Corps, 82d Airborne, 7th ID, 7th Special Forces Group, 75th Ranger Regiment, two battalions of the 5th ID, 193d Infantry Brigade, and several military police units comprised the bulk of Army troops committed to the fight.

The 1138th Military Police Company of the Missouri Army National Guard was the first activated unit on the ground. The command was already in Panama conducting its two-week annual training when it received federal activation orders and instructions to establish a detention camp for captured Panamanian forces. They organized a detention compound without attracting attention and were ready to receive prisoners before the fight began.

Superior American forces quickly suppressed organized opposition from the Panamanian forces. Endara was immediately sworn in as President. In a supporting operation titled "Nifty Package," US Navy Seals destroyed Noriega's jet as it was still on the tarmac and sank his boat while it was tied to the dock. Noriega hid in the Holy See Diplomatic Mission in Panama City, having no means to escape the country. American forces surrounded the mission and commenced the playing of rock music, which they knew Noriega despised. One of the songs played repeatedly was the Clash version of *I Fought the Law and the Law Won*.

SecDef Cheney, Vuono, and I were immediately on a flight to Panama once Thurman received the green light. Noriega was still cowering inside the Holy See. Some skirmishes were still going on, but technically, the battle was over. Panamanians freed from years of tyranny were doing an outstanding job of ridding their country of Noriega's holdouts. It was important for the SecDef and the Chief to receive command briefs. It was equally important for me to talk to the sergeants and junior troops. Any delays in getting to the warriors would result in critical details being lost. I needed as much firsthand information as possible for future discussions with corps and division command sergeants major. The best news I received was that radio communications worked correctly. We avoided the 1984 fiasco of Grenada, when failure in radio communications resulted in a resourceful US Marine calling back to Camp Lejeune, North Carolina on a pay phone to coordinate naval gun support.

We spent Christmas 1989 where we belonged, with service members who had just delivered an overwhelming victory for their country. On December 26th, Cheney, Vuono, and I were on a plane back to the United States. On that flight, my feathers got ruffled, and I let loose when a Navy captain on Cheney's staff made a comment when he should have exercised his right to remain silent. With Vuono and me present, the captain told Cheney, "I told all of you that you don't need heavy forces to fight a war." He claimed that we could fight and win with just C–130s, artillery, air, and naval gun fire. I turned loose on him. He and Cheney got an earful. To everything I said, Cheney acknowledged agreement. Vuono's continual nodding of approval let the navy captain know he would not pull rank to end this conversation. One fact is simple—that this non-combatant officer failed to understand: you cannot win a war unless you occupy the ground.

On January 3d, just nineteen days after his sword-waving declaration of war, a very exhausted Noriega surrendered to the Americans. He was quickly taken out of Panama to ensure the population knew his reign of terror was over. Noriega was subsequently convicted in American, French, and Panamanian courts. He died in 2017.

A subsequent CBS News poll found 92% of Panamanian adults approved of the invasion, with over 75% wishing we had not taken so long. US polling also revealed overwhelming approval. This operation suffered twenty-three US service members, and two civilians killed. The Panamanians suffered 314 soldiers, and 202 civilians killed. Indirectly, there was one more casualty. Shortly after the conclusion of JUST CAUSE, we learned that Thurman conducted the entire mission knowing he was suffering from incurable cancer. With the Panama invasion complete, his next request for retirement was approved by President Bush.

Vuono and I turned our attention back to daily operations. With the Department of Defense's Panama mission complete, it was now the State Department's job to work with the Panamanians in rebuilding their government.

Within all deployed commands, the review process began. It was important to secure all the facts before memories started twisting what happened. My starting points in these discussions were the notes I developed while talking to the NCOs and junior enlisted in Panama. One concern I addressed was how service members died by friendly fire. I never could accept fratricide, accidents, or friendly fire. If I die for my country, the enemy needs to kill me, not our own friendly fire. We had nineteen service members killed and even more injured from friendly fire. Most of it was stupid.

In addition to working on JUST CAUSE closure, I got to work on addressing the inability of Soviet tank crews to engage targets without first stabilizing. Joint conference calls with the sergeants major of TRADOC, the Armor School, and all of the divisions produced the same conclusion: if Soviet crews could not fire on the move, then neither could the less qualified troops of the nations that purchased Soviet tanks. It did not take much work for American crews to adjust to this new information. Our tacticians embraced knowing we could not be hit while Soviet tanks moved. Sharing this information with the Marine Corps produced predictable results. The Marines are always appreciative and make immediate adjustments when told how to better destroy the enemy. The Armor School had the contacts and took the lead in sharing the information with our allies, especially the British and West German. This knowledge was pure gold for all allied armor and antitank troops, and within a year we would be cashing in.

Focusing on future readiness, a problem with substandard officers and NCOs was ongoing and progressively worsening throughout the Army. Three interrelated factors were at play. First, there had not been a formal Army-wide selective retention review process since the early 1970s. Second was the negative effects of President Reagan's massive military expansion. Substandard officers and NCOs were getting promoted for being able to breathe, meeting their educational requirements, and not being convicted of felonies. For the Army, it meant the Active Component had shifted from promotion standards of "best qualified" to the Army Reserve's "fully qualified." Third was the growing lack of accountability and inflated evaluation reports being rendered throughout the Army. Receiving inflated evaluations along the way allowed these people to be continually passed on to someone else.

Problem individuals were not just undermining command effectiveness in their immediate presence. They were contaminating the development process for future soldiers and leaders. We did not allow the overwhelming and quick success of JUST CAUSE to cloud our assessment of the problem throughout the mainstream Army. We recognized that of the 27,684 total US military personnel involved, the Army's portion was primarily conducted with troops from the XVIII Airborne Corps, with other elite forces in support, and was not a reflection of the problem individuals who had worked their way through the system for almost two decades.

Vuono wanted this situation fixed before his end of tenure, which was coming up in a year and a half. With the Soviet threat significantly diminishing, we knew there was a window of opportunity to recalibrate the Army. We had to cut down the size of the Army and flush out the problem individuals. Ancient Greek philosopher Plato's assessment that "only the dead have seen the end of war" is accurate. There will always be a need for our nation to have an effective military. The keyword was effective, and we wanted to get it that way before the next conflict. Just as we did with the Army Pentagon staff, we had to trim down and clean up the Army ranks. The first place to start was the officer corps.

Vuono assigned the G–1 primary lead and the G–3 in support. We knew there would be those who played the system to their own advantage, no matter what we did. I was adamant that we should never again allow RIFed officers to become NCOs. That undermined NCO development, professionalism, and advancement in the 1960s and '70s. It also implied that NCOs were of a lesser quality where unsuccessful officers would be in their element. Vuono, Sullivan, and I had such a close relationship that they were not offended by my outspoken position on this subject. Instead, they backed me up with the rest of the senior Pentagon staff. They also knew I was not speaking only for myself. Word was returning from the command sergeants major throughout the Army that dumping poorly performing officers into the NCO Corps must never happen again.

Our goal was to get the details worked out and executed in 1990. Unfortunately, as we were developing the means for effectively resizing the ranks, retaining the best, and flushing out the worst, Iraq's Saddam Hussein invaded Kuwait. What became known as the Selective Retention Plan had to be delayed. Our nation had a high-intensity conflict to fight. The US Army was going to be the primary force provider.

We immediately accomplished getting as many substandard active component officers as possible away from positions where they would be leading troops into combat and could have any negative impact on combat operations. The problem had advanced high enough in the ranks that, in one case, we had to remove a deputy division commander from his position. I did not discover until years later that instead of completely removing this brigadier from the Army, the General Officer Management Office took the easy road and transferred him to ROTC.

A lot of people did not fully understand the threat that Iraq was imposing on the rest of the world. Saddam's Baath Party was founded to unite all the land formerly known as Mesopotamia and the Arabian Peninsula under one ruling government. Saddam had made known his intention to annex Kuwait since his seizure of the Iraqi government in 1979. Adding to this were Saddam's unpaid loans from Kuwait and Saudi Arabia of $14 billion and $20 billion, respectively. The successful annexation of both countries would have negated those loans and made him the supreme ruler of a united Arab empire. Saddam had a large standing army that the Soviet Union had equipped. Unfortunately, the Kuwaiti government had not taken Saddam's threat seriously. Its military consisted of 14,000 soldiers and 2,200 airmen. For a nation pulling in as much revenue as Kuwait with a warlord neighbor threatening invasion for over three decades, there was no excuse for such a weak military. The Saudi Arabian military was also in no condition to deal with an Iraqi invasion.

United States Central Command (CENTCOM) Commander General Herbert N. Schwarzkopf Jr. had recognized the possibility of a Saddam invasion, not only into Kuwait, but into the entire region south of Iraq. Earlier that year, 1990, Schwarzkopf ordered and oversaw the development of Operations Plan 1002-90, titled "Defense of the Arabian Peninsula." One week before the Iraqi invasion, CENTCOM had just completed an exercise of the plan.

Schwarzkopf was an interesting individual. He knew the Middle East better than any other general in the military at the time. He had lived in Iran multiple times. His father, also an American officer, was stationed in Tehran during the days of pro-American Shah Mohammad Reza Pahlavi. With an IQ of 168, Schwarzkopf had earned his master's degree in engineering. In Vietnam, he earned three Silver Stars and two Purple Hearts. While being an advisor to the South Vietnamese Airborne Division, he had been wounded four times by small arms fire in just one engagement. Yet, he refused medical evacuation until the mission was accomplished. Schwarzkopf had earned two nicknames, "Stormin' Norman" and "The Bear." The latter referred to his massive size, supported by his 6'3" frame. He was the right person in the right place at the right time. There was one downside. Schwarzkopf was a good tactician, but he seldom mentioned an individual soldier. That irritated a lot of us.

The Persian Gulf War consisted of four distinctive phases: Saddam's invasion of Kuwait, coalition deployment, preparation of the battlefield, and coalition assault into Kuwait and Iraq. This heavy operation involved all four branches of the US military and allied forces worldwide.

The first phase began on August 2, 1990. Saddam's forces hit Kuwait hard and within two days controlled the entire country. As the aggressors regrouped, the world realized the Saudi Arabian military was no better than Kuwait's. The Saudi government had been taking in enough revenue from the oil industry that the country could have afforded one of the best-trained and equipped militaries in the world. Instead, the royal family had spent most of the government's wealth on themselves. Saudi Arabia had little ability to counter Saddam's forces. On August 7th, Saudi King Fahd and President Bush came to an agreement on a US-led defense of his empire.

In the five days between Saddam's invasion and the Saudi King's approval to bring allied forces into his country, there was a flurry of activity in the Pentagon and throughout all American armed forces. We knew we were going to war. Personnel, training, and logistics were being examined and enhanced within all Army units. In divisions, the corps, the FORSCOM, the headquarters, and the Department of the Army, all focus was on fulfilling our responsibilities to support CENTCOM's execution of its "Defense of the Arabian Peninsula" plan.

If we had done nothing, Saddam would have taken over the entire Arabian Peninsula. Someone with Saddam's mindset, if allowed to go unchallenged, would regroup and continue his aggression as soon as possible. The Emirates, Yemen, and other countries nearby would have fallen. Once his power was consolidated, Saddam would have turned his attention to Jordan, Syria, and Israel. He had already sent several hit squads into Syria with the mission to assassinate President Hafez al-Assad. If successful, Saddam would not only have controlled most of the oil in the region, but his domain would have extended from the Persian Gulf to the Suez Canal in the southern sector and from Iran to the Mediterranean Sea in the northern region. Going one-on-one, he never could have defeated Israel but would have made life there a living hell. Turkey would likely have been left alone, but the fine balance of Middle East power between Iraq and Iran would have been shattered.

The second phase, deployment, began on August 20th with the arrival of the 82d Airborne Division in Saudi Arabia. They posted themselves in the desert and prepared to go head-to-head with Iraqi forces. Days later, other coalition forces started entering the country. In what was called Operation DESERT SHIELD, 500,000 US and 250,000 troops from other nations were deployed into Saudi Arabia. Not since the Normandy D-Day invasion in 1944 had so many forces landed so quickly into a foreign country.

If Saddam had kept rolling before the 82d arrived, it would be a different world. Instead, Saddam allowed us to have secure ports, airfields, and seaports where we could receive equipment. If the Iraqi forces had occupied those forces, it would have been a very complicated war that would have been difficult to overcome and win. Saddam's decisions to invade Kuwait and pause from going further cost him dearly. During those early days, the 82d was not there completely alone. The US Air Force was ready to drop any Iraqi plane that entered Saudi Arabian skies or bomb Iraqi army columns had they advanced toward the 82d or into Saudi Arabia. Additional US Army and Marine forces were quickly mobilized for deployment. Six US Navy carrier fleets were entering the fight.

Just as Schwarzkopf was the right person in the position he held, the same was true of Powell. He expanded the standards of the Weinberger Doctrine to the point that it was becoming his own. Each of his eight standards required an affirmative answer before engaging in war: Has American national security been threatened? Is there a clear, attainable objective? Has an honest risk and cost analysis been completed? Are nonviolent options no longer available? What is the exit strategy to avoid a war that does not end? What are the consequences of allied action? Do we have the support of the American people? Do we have broad international support? No decision to jump was going to be made if any of those doctrine elements was not satisfied.

Powell and the four chiefs of staff recognized that Schwarzkopf was the combatant commander and did everything they could to support him. I observed the respect the chiefs had for each other. Under Powell's direction, the chiefs worked together to plan a way ahead for everything that was developing. Everyone knew the Army and Marine Corps would do the ground fighting. Navy and Air Force missions were getting them into Saudi Arabia, long-range target engagement during the preparatory phase, and providing overhead support during the ground phase.

All of us in the senior officer and senior NCO ranks knew that, despite our own combat experience, this was going to be an entirely different war. Combat in the Middle East desert, against the fourth largest army in the world, was going to be no Vietnam, Grenada, and Panama. We had learned from mistakes in our past but were determined not to make new ones by refighting our last major conflict. The years of Army-wide desert training at Fort Irwin's National Training Center (NTC) in California were about to pay off.

Among the demands upon me personally were unbelievable hours. While Vuono engaged subordinate generals, I worked with the command sergeants major. In support of the mission, I was in constant movement, visiting with commands preparing for deployment. The working relationship established with the command sergeants major at the beginning of my term was paying off.

The sergeants major knew that my visits to their commands were not inspections or trying to find fault with them, but to learn what they needed and report back to the Chief. Waiting to address shortfalls until battalions arrived at the embarkation points would have been too late. Vuono needed to know both the good and the bad. The sergeants major ensured their respective commanders knew my visits meant Army-level support in responding to their problems.

Down to company level, each deploying command had to be combat effective. All personnel, training, and logistical requirements had to be met. Commanders had nothing to fear if they had properly completed their quarterly unit status reports. Concerning the Regular Army, this was not a problem. There were, however, cases within the Army Reserve and National Guard of reported inflation and manipulation. More direct—some commanders had been flat out lying.

While visiting the NTC, I observed a National Guard maneuver brigade being evaluated for its ability to deploy into a combat zone and provide leadership to subordinate commands. This command was so dysfunctional that it became obvious that its officers and NCOs never took discipline and training seriously. It would have been a tragedy to allow this brigade to waste aircraft space, send it into combat, and further trust it to lead battalions. Theirs were problems that could not be overcome in our time available. Even then, the first step in making this brigade command combat effective would have required a complete transfusion of new leaders into command. Despite the glowing reports it had done on itself prior to activation, the command had to be stood down.

The biggest problem we had to overcome was moving equipment from forward storage units in Europe and the continental United States to the Middle East. We discovered early that our ability to ship it, to roll our heavy force on and off ships, would be a difficult, but not insurmountable task. The coordination between the soldiers at Fort Bragg and the airmen at co-located Pope Air Force Base was impressive. I personally watched the 82d Airborne loading up for their departure. Meanwhile, Fort Hood was preparing the tanks for transport. This was the first time I had witnessed soldiers Army-wide preparing to execute combat missions. As for the soldiers, all they wanted to do was deploy into combat. During the staging process, that is all the soldiers talked about. Continually, I was asked, "Hey, when are we going to be able to go fight?" They were not bragging. They just wanted to use their training and equipment to prove they could fight and win a war for our country.

Talking to the soldiers and seeing their work made it easier for Vuono and me on those many trips over to Capitol Hill. Both during committee hearings and one-on-one conversations with congressional representatives, the questions always concerned command readiness and troop morale. Having been on the ground with those troops provided legitimacy to my affirmation that our nation would be successful. One question we did not have any problems with was the legitimacy of the coming war. Powell's and Bush's confirmation of the eight Weinberger-Powell Doctrine affirmatives took care of that.

Phase three came with Desert Storm's artillery and aerial campaign commencing on January 17, 1991. By then, we knew the troops on the ground could handle any offensive that Saddam could try to throw our way. His troops tried one poorly executed attack and were whipped by coalition forces. Concurrent with the artillery and aerial campaign, coalition troops continued to fine-tune their skills. Logistic operations continued to arrive in theaters in preparation for our going on the offensive. Unexpected anticipation caused us a little consternation. We did not know what we would be getting into once the ground war started. We prepared for worst-case scenarios. We were going to take down the fourth-largest army in the world, and Saddam had a history of using weapons of mass destruction.

One worst-case scenario that never materialized concerned medical units. Field hospitals were set up and equipped to receive mass casualties. The ground offensive was so brilliant that casualties were limited, at least on the coalition side. As in all combat operations, it is better to be prepared for the worst. Adapting to a better outcome is easy. Failure to have those medical units in place would have been criminal if casualties exceeded our capabilities, especially if Saddam used chemical or biological weapons.

The ground engagement was delayed until allied forces were ready. Generals Powell and Schwarzkopf caught heat because the media wanted things to jump off more quickly. Sometimes the media got out of hand. All those "military experts" thought they knew how to fight a war like one that had never been fought before. They did not fully understand how we had to prepare the battlefield. Many of those press people thought we needed to rush in there and start committing suicide. That is not the way it is done. Both generals weathered the storm. They were sticking to Army doctrine—destroy the enemy by aerial and indirect fire bombardment.

Our soldiers were getting beefed up while the enemy was getting torn up. One prime example of how this strategy works was VII Corps Commander, Lieutenant General Frederick M. Franks Jr. As part of his battlefield preparation, Franks was examining a situation map developed by his intelligence and operations staff. Referring to an Iraqi command positioned in his primary route of future advancement, he stated, "I want that brigade to go away." That brigade was pounded into nonexistence between his own artillery and Army aviation assets, with further support provided by the US Air Force. Saddam could have sent in more troops and equipment to replace that brigade. Nothing would have pleased Franks more. Whatever showed up would also have been annihilated before the ground engagement began. Franks was a good tactical officer, got things done, and was on a roll. The fact that Saddam did not send in more troops provided us with the intelligence we needed that Iraqi personnel and equipment resources were stretched across the front line.

Saddam had placed his most loyal Republican Guard forces behind the poorly trained regulars. Front-line Iraqi soldiers knew that if they attempted to retreat or desert, they would be shot. The problem was made worse by the fact that Republican Guard soldiers were Sunni Muslims. Most of the regulars on the line were Shia, whom the Republican Guard soldiers were willing to shoot without hesitation. There were only two ways these poorly trained soldiers could leave the battlefield: as a casualty or by surrender. There was a lot of surrendering among the Iraqi soldiers. One group even raised the white flag and surrendered to an Apache helicopter. Hovering above, the pilot escorted the Iraqis to the American lines, where they were taken into custody, processed as prisoners of war (POWs), and provided the first complete meal they had received in weeks.

There was one disturbing situation we did not realize until the fight was over. Percentages of Shia Arabs brought to our POW camps by Sunni coalition units were far lower than those brought by Americans, British, and every other non-Arab nation. We were fighting to free Kuwait from Saddam. The Sunnis were continuing with their religious war that started at the Battle of Karbala in 680 AD. The danger of allowing either Sunnis or Shias to have the authority to determine the lives of each other is one lesson that should never have been forgotten. Twelve years later, with the US-led invasion to topple Saddam, it was forgotten with disastrous results.

Attempting to make this an "Arab versus the World" war, Saddam ordered the launching of Scud missiles at Israel. Few of the missiles did any damage. In one case, the Israelis congratulated Saddam on being able to strike the desert. Among the first targets taken out by the US Air Force were Saddam's nuclear weapons al-Tuwaitha facility, built after the Israeli strike on and decimation of the original facility twenty years earlier. This American strike permanently ended Saddam's attempts to build a nuclear weapon. Once the Navy's carrier groups arrived, Saddam was getting an overwhelming dose of American airpower. Instead of trying to put his planes in the air to dogfight with American aviation, Saddam commenced trying to hide many of his planes by burying them in the ground. That was ridiculous because all Saddam accomplished was destroying the jet engines with gravel and sand. If he wanted to destroy the planes, he should have been courteous enough to send them into combat and let US Navy, Marine, and Air Force pilots have the satisfaction of shooting them down. It was cowardly of Saddam to deny our pilots the opportunity to have those combat kills.

During the latter half of the air campaign, I took a trip to Saudi Arabia. Continuing to train and perform equipment maintenance kept soldiers very busy. Sitting around and waiting devastates morale while producing rumors, uncertainty, and mischief. Once again, the troops kept asking when they could go and fight. I assured them they would soon have the chance to engage the enemy. In the meantime, we needed to give the aviation and indirect fire components time to increase our friendly-to-enemy casualty rate. Upon receiving the order to attack, they focused on destroying the enemy. They were not worried about going home, getting mail, or anything else.

As part of his deception operation, Schwarzkopf intentionally let it slip to the press that US Marines were offshore preparing for an amphibious landing. The Iraqis dumped eleven million barrels of crude oil into the Persian Gulf to inhibit the landing and endanger allied ships. Saddam kept troops on the coast to repel an attack that was not coming. Only a few of us knew in advance that Schwarzkopf would exercise a "Hail Mary" maneuver on the Iraqi military. Keeping the Iraqis occupied directly in front of the forces, Schwarzkopf was preparing to have his divisions circle to the west and come in from the flanks and rear of the Iraqis. Patton called this tactic "grab them by the nose and kick them in the pants."

By mid-February 1991, Schwarzkopf and the Joint Chiefs reported to SecDef Cheney and President Bush that all ground forces were ready for the assault. Bush sent an ultimatum to Saddam to either pull his forces out of Kuwait by noon in February 23d or face the consequences. Believing the attack would be delayed, Saddam ignored the ultimatum. Before daylight broke on the 24th, the thirty-nine days of aerial and artillery preparation of the battlefield ended. The ground invasion was on.

Abrams was right: "Generals may plan battles, but they cannot advance very far without soldiers." Our troops came out like coiled springs. As we intended, Iraqi troops had anticipated the invasion from the south and instead received the main thrust from the undefended west.

I was with Vuono and his staff at the Pentagon when the ground assault started. There was nothing we could do but watch. Even Schwarzkopf at his headquarters could only monitor incoming reports. It was junior enlisted, NCOs, and young officers—engaging the enemy and moving exactly on the routes of advance assigned to them—who were now winning this war. Battalion and brigade commanders were supervising the progress of subordinate commands. Division and corps commanders were ensuring that any additional support and information needed were made available to the frontline troops.

Many of our own airborne commanders had wanted to enter the ground campaign with a jump. Jumps are usually a systematic way for advancing ground troops to punch through the enemy lines and relieve the airborne troops. For four reasons, Schwarzkopf refused. Using airborne soldiers in the heavy environment that he and his corps commanders were constructing would have produced many casualties that would not have happened in a light environment. The speed at which he programmed and planned the fight to proceed would have armor and mechanized infantry units at the same exact locations the airborne soldiers would have taken a few hours earlier, and at far less risk. Having nothing in front of his tanks and Bradley fighting vehicles except enemy troops greatly reduced the risk of friendly fire. Schwarzkopf recognized that a lot of hidden Iraqi air defense systems would still be functional when the ground war started.

He used the 82d and 101st Airborne divisions in a fantastic way, transporting them by trucks and disembarking them as necessary. Tanks should never deploy very far without infantry. At the speed at which those tanks moved into Iraq, soldiers on foot would never have been able to keep up. Our armor, artillery, and infantry troops moved quickly and with such precision that the only chance for Iraqi soldiers' survival was to raise white flags. There was plenty of that. Many Iraqi regulars had been "press-ganged" into the military and had no idea of how to use their equipment. Even the ones carrying only rifles had minimal skills in lining up the front and rear sights. Iraqi armor crews were no match against their American counterparts. Once they had stabilized their T–72 tanks, it took the Iraqis a lot longer to fire than the Soviet crews like Vuono and I had witnessed during our trip to Moscow. With great precision, our crews hit targets without decreasing maneuver speed. It was a one-sided fight.

Twenty-six American soldiers were killed and over one hundred were hospitalized during the one-hundred-hour invasion when an Iraqi Scud missile struck a warehouse in Dhahran, Saudi Arabia. The building served as living quarters for the US Army Reserve's 475th Quartermaster Company from Farrell, Pennsylvania. Eyewitnesses claim the Scud was passing over the top of the warehouse when an American Patriot missile hit it. Subsequent investigation could not confirm a Patriot was fired. Even if it is true, criticism of the Patriot crew was unwarranted. That Scud was going to hit somewhere in Dhahran. The Patriot crew had no way of predetermining exactly what would be on the ground below when the Patriot engaged the Scud. If a Scud missile were to land because a Patriot crew knew it was coming in and elected to do nothing, they would have been held accountable.

After one hundred hours, President Bush called off the assault. The mission to remove Iraq from Kuwait was complete. Once the fight was over, Vuono and I toured Kuwait and the southern portion of Iraq still under coalition control. Our four-hour helicopter trip included a run up the "Highway of Death." There were thousands upon thousands of vehicles, including Mercedes-Benz privately owned vehicles. It's hard to explain the number of destroyed armored vehicles that were out on that desert that once belonged to Saddam's army. It had to be seen to be believed.

Another unforgettable scene was the number of Kuwaiti oil rigs the Iraqi military ignited during their retreat. Day and night, over 600 infernos were producing black smoke everywhere. The massive smoke had to have negative long-term effects on our soldiers. Conservative estimates predicted it would take years to extinguish those flames. "Hell fighters," including American icon Paul N "Red" Adair, had those fires out in six months.

The allied coalition consisted of thirty-nine nations, including several Arab countries. Of the twenty-eight countries that contributed a total of 670,000 troops, 425,000 were from the United States. Iraq entered the conflict with 4,300 tanks, of which 3,700 were destroyed through the combination of coalition aerial bombardment, artillery shelling, and direct engagement fighting. The coalition sustained 292 deaths, against estimates of up to 100,000 Iraqi military deaths. 70,000 Iraqis were now in our POW camps. Our troops even took a few Soviet prisoners.

Vuono and I had three goals while visiting the Middle East. First was an eyes-on assessment of the war itself. Among the things we wanted to examine were destroyed Iraqi tanks, equipment, and anti-aircraft sites. The difference in what kind of shots destroyed the vehicles was easy to identify. The years of America's military advances had paid off. The second was to receive after-action briefings from all levels of command. We wanted to take back to the Pentagon staff the lessons of what happened, what went right, what went wrong, and what we needed to do to improve before the next fight. The third was to spend time with the ground troops. We could not visit all of them, but we wanted to see as many as possible. The soldiers knew what they had accomplished. They knew they went into the fight with the best possible equipment, had been thoroughly trained, and were led by commanders who would not put them at unnecessary risk. They had earned the right to express their pride in a well-done mission. Vuono and I wanted to hear it from them.

In a conflict as successful as Desert Storm, it is easy to be self-blinded. Good leaders must be careful not to let bad outcomes get inside their heads and affect their judgment to the point that they make future bad decisions. It is even more important that they do not let good outcomes affect their judgment, so they fail to recognize the need for improvement.

The US military had achieved its predetermined state, and now it was time to begin the exit strategy. Operation DESERT STORM yielded to

Operation DESERT FAREWELL. Unlike the days of Vietnam, American service members were returning to a nation that was welcoming them home.

CHAPTER 18
Epilogue

Following Desert Storm, Vuono and I spent our final five months in the Army working to bring units home, overseeing the consolidation of after-action reports into an Army-level analysis, examining all aspects of the Gulf War, ensuring major commands were getting the resources to reconstitute their units, and addressing soldier issues. Our visits to the bases were a combination of a farewell tour and developing lessons learned to bring back to the Pentagon. The workload remained intensive, but the pre-ground war anxiety about how everything would play out was gone. The years of rebuilding the Army, the intensive focus on training, and the raising of standards had paid off.

Ultimately, all fights come down to the warriors engaging the enemy. Our job was to prepare them with the right training, provide them with professional leadership, arm them with the right equipment, inform them with the right intelligence, and ensure coordination existed for the right fire support, whether it comes from indirect fire or overhead bombers. To the enemy, our warriors successfully got into their decision-making cycle and made the enemy react to us, not the other way around. The shortfalls of both Noriega and Saddam were so great that they were never able to overcome our strengths.

Vuono and I did not lead any troops in the low and high-intensity conflicts that occurred on our watch. Like our predecessors before us, since the days of Vietnam, we worked hard to build an All-Volunteer Army. There were a lot of bumps in the road as we moved along, but we started having equipment to enable the Army to overcome threats of the world and enemies of our nation. Against the Soviet military, we intended to take the battle to them and kick their butts before they could get across the border.

Wrapping up our mission as Chief of Staff and Sergeant Major of the Army was bittersweet for Vuono and me. We enjoyed what we were doing and were able to see the results of not just our work but that of all our predecessors. But we both had families who deserved attention that we could not provide them in the previous year and a half.

It was time for us to close that chapter in our lives. Stepping out of our uniforms did not mean we ended our loyalty and service to our nation. That will remain as long as we live. It was time for new leaders, who once followed us, to take the helm and lead the Army forward even more. When we transitioned our posts to our replacements in June 1991, we knew we had fulfilled President Harry Truman's adage, "All power is temporary; leave it in the best condition possible."

APPENDIX
Professional Soldiering

One of the legacies of the NCO Corps since the days of Alexander the Great is its ability to pass knowledge through the ranks. Those who answered the sound of the guns following Lexington and Concord were not completely green troops. Many warriors of the American Revolution had received their baptism by fire in the French and Indian Wars.

Hopefully, this book adds to that legacy. In addition to the previous experiences and lessons shared, the following observations and assessments concerning professional soldiering are provided.

Have the willingness to re-evaluate your own beliefs and conclusions.

During the ground offensive of Desert Storm, I wished President George H.W. Bush would delay calling the cease-fire and allow the coalition to either capture or kill Saddam. Giving credit where it is due, President Bush recognized the entire Middle East environment and was thinking long-term. He remained in compliance with both the Weinberger-Powell Doctrine and the original intent of engaging Saddam only to force him out of Kuwait.

International support was recruited for that specified mission. To go further would have violated the agreement with coalition partners. President Bush also recognized another concern. The balance of power between Iraq and Iran would have been destroyed if Saddam had been taken down. The result would have been total regional anarchy.

President George H. W. Bush was right in his decision.

A major, but highly ignored, part of leadership is listening.

One time, as Sergeant Major of the Army, I received a professionally organized and well-presented briefing from a soldier. Partway through the briefing, he stopped and asked me if everything was all right. I assured him his briefing was on target, and he was doing a fine job. He responded, "Sergeant Major, you are the first to not interrupt me while I was talking." I replied, "I have to listen to what you are saying to understand what it is you want me to know." The soldier broke out with a big grin and continued. His presentation was so thorough that when he finished, we had more discussions rather than questions and answers.

Too many supervisors spend more time talking than they do listening. Such behavior is counterproductive. Supervisors cannot learn when they are the ones doing most of the talking. Let subordinates know their knowledge is respected. When a dedicated soldier feels something is important enough to say, it is important enough for the senior to listen to it. This helps build their confidence and communication skills. There are those subordinates, even peers and seniors, who want to make noise about nothing. They are not hard to identify. Being a good leader requires the ability to listen. It takes a little while to learn how to listen, but it must be done. Some take longer than others. Some never learn it at all and never develop to their full potential.

Allowing subordinates to ask questions is also important. Sometimes the questions will be amusing. While visiting a battalion during maneuver training a junior enlisted soldier came up and said, "Sergeant Major, may I ask you a question?" I replied, "You can ask me anything you want." She proceeded, "Are you really the Sergeant Major of the Army?" I assured her this was true. She then stated, "But you seem too small to be the Sergeant Major of the Army." Figuring she must have been expecting someone looking like an all-star wrestler, I said, "Dynamite comes in small packages." Her eyes grew wide, and she said, "You're dynamite?" Now enjoying the conversation, I responded, "I'm full of dynamite!" After we finished the conversation and walked away with her battalion command sergeant major, I overheard other NCOs reprimanding her. I walked back and told them to stop. She had asked for and received permission. I then presented her with one of my Sergeant Major of the Army performance recognition coins because she had the courage to ask the question.

Understand the interconnectivity of the three pillars of training and soldier development.

The three pillars of training and soldier development are institutional courses, unit training, and individual self-development. When applied separately, they will never achieve the maximum possible results. Of all the people I have ever known, Sergeant Franco was the best at interconnecting those three pillars. Before his soldiers were sent to advanced formal schooling, he had graduates of those courses provide pre-schooling. People learn from repetition.

By preparing and teaching those advanced classes, our unit instructors had to refamiliarize themselves with what they had been taught. The best way to lock in knowledge on a specific subject is to teach it. Because of Sergeant Franco's initiative, his troops arrived at the school already knowing much of what would be presented. That allowed them to concentrate more on the detailed information while other students were trying to learn everything from scratch.

Individual self-development requires self-motivation. This includes the classroom, after-hours study, and especially applying oneself in daily unit missions. No one will ever improve themselves by shirking responsibilities and opportunities to learn. Here, too, the importance of institutional and unit training is important. No student should ever have to attend training that is not effectively developed and professionally presented. Check-the-box courses are a waste of time.

Leadership must be conducted from the front.

The best leaders are not only tough, but they also keep their subordinates informed, are not afraid to put themselves at personal risk, and look out for the long-term welfare of their troops.

In Vietnam, then-Colonel Hank Emerson set the standard for walking the walk. He would always do something to motivate us every time we went on a difficult mission. Prior to the 1966 Dak To Mission, he stood on top of a Jeep and explained to us the purpose of what we would be doing. He then motivated all of us by explaining that we needed to go out, do a good job, and bring all our troops back alive.

When Colonel Emerson had multiple commands deployed in the field during his recondo/checkerboard operations, he would be monitoring the entire battle sector from his helicopter. He did not micromanage anyone but rather worked closely with the battalion commanders to make sure they could respond to the fluid situation. Colonel Emerson had an artillery battalion dedicated to his brigade. Any unit with a serious problem knew artillery support was immediately available. From the ground, we would request indirect fire support. Emerson would demand it from the air if we did not get it fast enough. His presence also ensured his subordinate units did not get into situations where friendly fire could occur.

We had to carry combat rations in our backpacks on our long-range missions. Every five days we would get resupplied. Included with the resupply, Emerson had the dining facility prepare something special for us, like steak sandwiches. He was not a soft commander. He was tough and expected his soldiers to be just as tough and to know what they were doing. Emerson found the right combination that allowed him to take care of his people and win battles.

Later as Lieutenant General Emerson, he commanded the XVIII Airborne Corps. He went head-to-head with the military-industrial complex and senior Army leadership by exposing flaws in the first version of Bradley fighting vehicles. Time proved him right, and eventually the corrections were made. The Bradley was turned into an outstanding piece of equipment, but not before Emerson found himself on the outside of the good-ole-boy network and subsequently pressured to retire. There can be no doubt he would have raised the red flag all over again if it meant taking care of his soldiers.

Know how to lead people and develop teams.

Trying to always be the nice supervisor ultimately never works. That will just set the stage for those who wish to take advantage of the situation. It also undermines the dedication of those who want to do it right.

Motivating subordinates to do what's right starts with setting the example. All people learn better by observation than they learn from what they are told. If the task is hard, the leader must do it first or have an established record of doing it. Once a subordinate sees it can be done and is properly trained, then the leader's responsibility becomes making sure the subordinate performs it correctly. This is nothing more than establishing the tasks, identifying the conditions, and enforcing the standards. What works for developing individual subordinates also works for teams.

If a soldier is trying but not able to perform responsibilities to established standards, the soldier should receive additional training. Nonjudicial punishment and additional duty unrelated to the task do not constitute additional training. That training can involve evenings and weekends. The sergeant himself should be with the soldier doing additional training. The sergeant's job does not begin with reveille in the morning and end with retreat in the late afternoon. A professional sergeant is on duty twenty-four hours a day, seven days a week.

If soldiers are trained correctly and disciplined appropriately, we don't have to worry about them doing their jobs in peacetime or combat. In my thirty-two years in the Army, I recommended one soldier for an Article 15 (nonjudicial punishment without formal charges). I still kick myself for doing that. I never recommended a soldier for court-martial. Soldiers getting themselves into serious trouble, like driving while intoxicated, starting fights, property theft, abuse of alcohol, sexual assault, or other blatant violations of the UCMJ, are not recommended for judicial or nonjudicial punishment by their superiors. These soldiers have already recommended themselves for discipline by their conduct.

One thing that I have always despised is that seniors use physical force on their subordinates. This is immoral and illegal. In combat or a life-threatening situation, a supervisor may have to push or grab a subordinate to get him or her to move. But any supervisor who either boasts about "kicking a subordinate's butt" or lays a hand on a subordinate is entirely wrong in any military force. That person has also crossed the line in terms of self-recommendation for punishment under the UCMJ.

The importance of the relationship between Be, Know, Do.

Schools teach the concept of "Be, Know, and Do." It's leadership within commands, all the way from squad leaders to senior commanders, who need to make these concepts a visible and continuing reality. A lot of emphasis is placed on these three words, but we still have not gotten it right. Examining each in reverse order may help present the solution.

"Do" is actively working to make the right things happen. It can be applied everywhere. A good example is the motor pool in northern Germany. Everyone knew that the vehicles were not being properly maintained. The trash on the fence line and the sloppiness in parking the vehicles were indications and warnings that this was a slack unit. It was no surprise that maintenance on the vehicles was not being properly performed. Everyone knew what was happening. No one did anything about it until I took personal time away from the battalion NCOs and forced them to make it right.

That takes us to "Know"—technical and tactical proficiency. Knowledge is gained through studying, practical application, and repetition. This is where the three pillars of education play the most important part. Even if formal schooling is doing everything right, it is a wasted effort for soldiers to return to unprofessional command environments. The unit must also be aggressively engaged in creating an environment where soldiers are provided with training and encouragement.

"Be" is the inner character of the individual on which everything else is based. A leader must have a high threshold of moral ethics and must not seek personal gain at the expense of others, especially subordinates. Subordinates are quick to notice character flaws in a supervisor. They will follow that supervisor only because they have no choice, not out of trust and confidence. They also know a self-serving supervisor will throw them under the bus whenever something goes wrong.

Ignoring a problem does not make it go away.

Closely tied to be, know, and do are the four Cs: Courage, Candor, Competence, and Commitment. Courage is required not just on the battlefield, but in everyday operations. If something is wrong, courage allows one to address the situation. Candor requires frankness. Rudeness and frankness are two different things. Rudeness is the product of a prima donna and adds no value to anything. Competence goes back to "Know," which comes out of learning and self-application. A leader has several commitments: to seniors, to mission, to subordinates, to family, to the truth, to our nation, just to name the most obvious. Commitment to oneself is necessary for personal growth, development, and long-term success. Self-commitment should not be at the expense of others—that's selfishness.

By exercising the four Cs, General Vuono and I had a close relationship. I always presented myself as being loyal to the Chief, and Vuono was worthy of all the respect I afforded him. Behind closed doors, we had some very focused and direct conversations. He always welcomed my opinions and stances. He knew where I was coming from and knew the value of the information he was being provided.

Some senior officers were upset with my aggressiveness against allowing RIFed officers to be dumped into the NCO Corps. I had no problem with substandard officers, knowing I was a major reason they were not going to get a golden parachute, allowing them a soft landing to their retirement. What was important was Vuono's support for my position. He knew my courage and candor were anchored in my competence and in my commitment to him, the NCO Corps, the Army, and national security.

When dealing with problem subordinates, ignoring or covering up their behavior or incompetence will only make it worse. Leaders must have the courage to look a problem subordinate in the eye and let them know their behavior is unacceptable.

Battalion is the level that can have the most positive impact on soldier development and success.

Battalion commanders and command sergeants major are the two leaders most critical in developing soldiers, which in turn determines the success of companies. No one should ever become a battalion commander or command sergeant major without having successfully served as a company commander or first sergeant. This is where they learn what does and does not work in running company-level units. When they move up to battalion, they bring those lessons with them, where they can mentor their subordinate leaders.

Brigades are focused on battle-space operations and are dependent on battalions to have the troops and equipment ready to respond. Operation JUST CAUSE was a low intensity conflict requiring a limited amount of engagement. Desert Shield/Storm allowed for extensive deployment and preparation of the battlefield. We must always be ready for a "come as you are" conflict. Service members must always be trained by task, conditions, and standards and be ready for quick deployment. Equipment must always be maintained in the best condition possible. Failure to do so is nothing short of negligence on the part of a battalion commander and command sergeant major.

Command sergeants major should never be just senior enlisted advisors to battalion commanders. They should be aggressively involved in directing and mentoring the first sergeants, ensuring soldiers are being paid on time, receiving proper administrative support, and being properly fed, properly trained, properly equipped, properly supervised, and, when necessary, properly disciplined.

Recognize the importance of families.

An expression in the 1800s was, "Bachelors make the best soldiers, all they have to lose is their loneliness." Except in combat or remote operations, in the American military, the daily Spartan life of soldiering no longer exists. While housing has been improved, there is still a need for improvement in financial support. No family of a service member should be dependent on receiving food stamps to make ends meet.

Margaret not only looked out for me, but also for everyone else. Subordinates have the habit of not telling their supervisors when something is wrong. From the spouses, Margaret learned the truth. That made my ability to be an effective leader much easier. If families were having a difficult time, living in poorly maintained housing, were subjected to spousal abuse, or anything else that was wrong, Margaret found it out. When she relayed it to me, it became my responsibility to resolve. Problems can grow like cancer, like the neglect and abuse suffered by those Korean children at the orphanage. Through Margaret's involvement, we were able to prevent a lot of bad situations from getting worse.

Also invited to the division level and higher sergeants major conferences when I was Sergeant Major of the Army were my seven predecessors. Having come for my first conference, Bill Wooldridge and his successor, George Dunaway, were sitting at a table in the post exchange food court when Karen Morrell and Margaret walked in. No stranger to them, Karen proceeded to their table, exchanged greetings, and introduced Margaret. Already having stood up, the gentlemen started to introduce themselves. With a big smile, Margaret cut in and said, "I know who you are," and referred to both by first name as she shook their hands. Margaret had seen and memorized the Sergeant Major of the Army's photographs and biographical summaries. With that, she won Wooldridge's heart, and forever after, he would tell people Margaret was the best thing that ever happened in my life. He was right. Except when in combat, she was always with me. Margaret was the reason I looked forward to coming home at the end of the day.

Without a family, it is likely I would not have toured Europe or enjoyed the communities around the bases. My family made me a better person and soldier. There is no reason the opportunity to have a family should be denied by any service member.

Have the courage to speak up when you know the truth.

While I was the 3d ID's command sergeant major in Germany, there was an Association of the United States Army convention in Europe where attendees could stay in a hotel for a low cost. The 8th ID sponsored the event and had the home court advantage for bringing the most attendees. In the spirit of competition, General Crowell and I determined the 3d ID was going to beat out the 8th on participation, which we did.

As representatives of our command, the soldier and the NCO of the year were on orders to go to this convention. Crowell instructed me to escort them to the convention in one of our staff cars. Someone had reported to the military police provost marshal that I had used the vehicle for personal reasons. The provost, who was a member of the division staff, could have resolved the entire issue with a simple question to either the general or the division chief of staff. Instead, the provost referred the matter to the Department of the Army Criminal Investigation Division (CID), which in turn launched an investigation into me for misuse of a military vehicle.

I asked the CG's executive officer what was going on because he was present when Crowell approved my use of the vehicle, but instead of saying anything, he went silent. No way would I ever go to war with someone like that. His behavior was unacceptable for someone either in a position of authority or in possession of information. When CID came to talk to me, the division chief of staff informed the agents to remove themselves from the command section, that the investigation was over, and he was assuming responsibility. He further told the agents that if they had a problem with that, they should take it up with the provost or the commanding general. In my three-plus decades in the Army, I have made some mistakes. Misuse of military vehicles was not among them. Vehicles cost a lot of taxpayer money and must be used carefully and legally.

Understand all the facts before criticizing anyone.

As division command sergeant major, I would always observe when battalions receive an unannounced alert to deploy into the field. A medical unit was evaluated hard because they had vehicles in the motor pool without enough drivers. Senior leaders were sitting in the general's office criticizing the battalion commander, and with me being an old country boy, I told them to stop and look before they made a judgment about people. I already knew what the problem was because the battalion sergeant major had told me they had a high percentage of pregnant female soldiers in the command who could not drive the vehicles. In writing, the battalion commander had continually been reporting the problem and requesting assistance. It was not the battalion commander's fault; it was the fault of his seniors who ignored his written requests for assistance.

Transferring soldiers between commands is very easy; it could have been done immediately when the battalion commander asked for assistance. Some of the finest soldiers in the US Army are women, but things happen. When I was Sergeant Major of the Army, we had more men on light duty than women, even including pregnancies.

Looking for the best in people usually produces successful results.

One of the first challenges Sergeant Franco assigned to me after I was advanced to squad leader was consolidating all the platoon's weaker soldiers under my control. He made it clear that in putting them together within the same squad, they could receive special development away from the peer pressure of others. They were all good people who wanted to be among the best. They just needed additional attention and training. Getting the job done required applying skills I had gained in the Army and on the farm. The wiser the land is worked, the better the crop yield.

Sergeant Franco's additional task was to turn those soldiers into the best squad in the platoon. Just as my father assigned work to all of us boys, we did the same thing in the squad. On the first day, I informed them that they were "not going to be the duds anymore." They were "going to be soldiers, the best in the platoon." They dedicated themselves to the challenges placed on them by Franco and me. We trained hard.

As we were doing things, whether police call, latrine duty, or wall-locker inspections, I assigned individual soldiers to be the task leader. I would join the work detail. Soon, these soldiers noticed the change, saying, "Hey, something's different here." Perhaps because they were too tired at the end of the day, or because they decided to become the best soldiers possible, problems stopped. When the unit held competitions, our squad began monopolizing the awards. During the annual training test, that squad of so-called "dud" soldiers came in second best in the brigade, and third best in the division. Those soldiers always had it in them to win, they just needed the confidence and training to bring them to the top.

Recognize the importance of face-to-face communication.

There are three elements of communication: the person speaking, the person listening, and the situation. Electronic mail is a high-tech form of miscommunication. While it is easy to send communications through computer networks, there is a greater opportunity for the message not to be understood. Eye contact and the reading of facial expressions cannot be gained from electronic messaging.

Before computers and telephones, people had to go around and talk. Leaders still must do that with their subordinates. At all levels, seniors should not solely rely on messaging to communicate with subordinates. To subordinates, leaders must present themselves as part of the solution. If there is some way to help a soldier get better, do it. Praise is more constructive than criticism. Too many people in supervisory positions fail to understand that. When things are not right, address them. That, too, is part of the solution. Supervisors looking for ways to inflate their egos, intentionally finding fault in subordinates, is wrong.

In the 3d ID, I tried to visit every unit in the division as often as I could and to stay as long as possible. Before leaving, I visited with the local commander and command sergeant major. My observations and their concerns were thoroughly discussed. They knew my visits were going to mean resources and assistance as necessary, so they could better run their commands. If they were messing up, they were told that as well. I never told anything to the division commander that his subordinate commanders did not already know.

Not repeating the failure of the Soviet Union and its military.

The Soviets never really could have defeated the West. They lacked technology and certainly lacked depth in unit-level leadership. The Soviets failed to maximize the capabilities of their soldiers. In NATO countries, NCOs are the primary leaders of troops. They receive instructions from their officers and proceed with missions, working closely with their lieutenants and captains.

In the Soviet Army, and now the Russian Army, officers do the work of sergeants. Even before realizing Soviet T–72s needed to be stabilized before engaging targets. I knew the Soviets were no match against the Americans when the lieutenant colonel was leading drill and ceremonies. In all branches of the US military, that is an NCO's job. Furthermore, we encourage leadership and initiative at all levels, including junior enlisted.

American service members are the most unique in the world.

In combat, if an officer goes down, the senior NCO takes charge. If the senior NCO falls, the next in line will take the leadership reins, all the way down the ranks. They can be taken from the snow and ice in Alaska and a month later be in the jungles of Panama or the deserts of the Middle East. They will survive and they will function. If their equipment malfunctions and cannot be immediately replaced, they will either find a way to fix it or work around the problem. Using a telephone credit card, like one resourceful service member did during the Grenada invasion, to call in for additional firepower was a classic example. A Russian soldier would never do anything like that.

There is something magical about American service members. Part of it is our culture. The bigger part of it is that we are trained to adapt and encouraged to think outside the box. The difference is that Moscow's style of leadership is totally lock-step. In the American military, junior officers and NCOs can make decisions. But in the Soviet/Russian Army, it's all totally centralized.

Even though I have greatly referenced my time as a Ranger, one thing that must be clear is that no one specialty of soldier is better than another. That includes Rangers. As a rule, Rangers are better trained and, by the nature of their environment, have better discipline. Ever since the days of Lieutenant Colonel Robert Rogers in the French and Indian War, Rangers have pushed the envelope of their capabilities. That does not afford them exclusive rights to excellence and professionalism. All soldiers have the same opportunity within their specialties to "Be All You Can Be." This motto applies to every soldier from private to general. It applies to every team, section, and command throughout the Army.

That was the smartest recruiting slogan the Army ever developed, and we were fortunate to have it throughout most of my military career. In my view, it never should have been replaced. It encourages soldiers to work hard, study hard, and earn the pride they should always have in themselves.

Thinking something is owed.

We must be careful when we think people owe us something when they really do not. In my opinion, there are three classes of people: those who will always excel, the average group who apply themselves to a 75% standard and mostly stay out of trouble, and the class who are not motivated to achieve at least 75%. This last group I called the "My name is Jimmy. How much can you give me?" category.

The goal must be to get the lower groups to move up a step. It is not easy to do, but it can be done. Role models are especially important to individual development. Before leaders expect their troops to achieve a high standard, they had better make sure they are doing it themselves. Leaders who create professional environments are going to encourage their subordinates to be professional. Unprofessional environments will most certainly produce unprofessional soldiers. The adage is true: "If you talk the talk, then walk the walk."

Earn It.

The gifts of life and the ability to serve our nation as well as humanity require effort, commitment, and self-sacrifice. Fulfillment cannot be achieved by inactivity and expecting everything to be done with little to no effort, especially in our officer and NCO promotion system. Years ago, to be promoted, we were required to have time in grade and service, be specialty qualified, and have demonstrated ability to function at the next level. Supervisory promotions based only on longevity and specialty qualifications are wrong. Unearned promotions are a serious problem at any level, but even more so at ground zero, where training, development, and the immediate future of our service members are dependent on their unit leaders.

Ironically, this comes at a time when we have the best leadership development schools. The problem is not in the academic environment; schools can teach the concept of leadership, and they can teach the required skills and proficiencies necessary to survive on the battlefield. It is the individual service member and the unit that must set the standards for success. This takes us to the concept of "Earn It." People, especially those in the military, cannot expect something to just be given to them. Increased responsibility or authority that is given, but not prepared for by earning it, will result in failure. Military failure, whether on the battlefield or in support elements during battle, results in the increased use of body bags.

Furthermore, if someone did not work for what they have received, then why should they work now? The reward for their past inactivity was advancement. Therefore, their learned behavior is to do nothing and get rewarded. They have no reason to change; they have achieved their personal self-serving desires without having to give of themselves. Meanwhile, the motivation and retention of those who practice the concept of "earn it" are negatively impacted.

At all levels, members of the armed forces have the opportunity to earn it and never stop earning it. The commander of the battle at Ia Drang Valley, Hal Moore, put it this way: "In any environment, there is something you can do to make it better, and something after that, and something after that." Whether it is self-development, team development, unit readiness, goal achievement, or any other worthwhile endeavor we accept, further improvement should always be the goal.

Moore's preparation for combat and his actions at Ia Drang were portrayed in the motion picture *We Were Soldiers*. When later asked if he prayed over the dead soldiers while still on the battlefield, as did Mel Gibson while playing his character, Moore responded that it did not happen because when the battle commenced, "The time for prayer was over; the time for fighting had started."

The same holds true for all preparation; when the fighting starts, preparation is past. The result will be determined by how ready we are. That includes coming with service members possessing the technical and tactical skills necessary not just to survive, but to achieve victory. They must have the spirit and toughness to go through the long fight. They must have confidence in themselves and their leaders. This cannot be achieved by individuals who have always been given the easy route and are supervised by people in positions of authority who have never earned it.

We have the greatest country in the history of the world, and we must do the same things our forefathers did, including defending it. We can defend it a lot better in our time of challenge if every step of the way we take the harder right, rather than the easier wrong. Earn It.

Acronyms

AAR:	After Action Review
AD:	Armored Division
AIT:	Advanced Individual Training
ARTEP	Army Training and Evaluation Program
BDU:	Battle Dress Uniform
BCT:	Basic Combat Training
CCC:	Civilian Conservation Corps
CFC:	Combined Forces Command (Korea)
CID	Criminal Investigation Division
CQ:	Charge of Quarters
DCG:	Deputy Commanding General
DMZ:	Demilitarized Zone
FORSCOM:	Forces Command
GED:	General Educational Development
ID:	Infantry Division
IG:	Inspector General
LOH:	Light Observation Helicopter
METL	Mission Essential Task Lists
MOS:	Military Occupational Specialty
MOUT:	Military Operations in Urban Terrain
MREs:	Meals Ready to Eat
NATO:	North Atlantic Treaty Organization
NCO:	Noncommissioned Officer
NVA:	North Vietnamese Army
NTC:	National Training Center
OCS:	Officer Candidate School
OPFOR:	Opposing Forces
OPSEC:	Operations Security
PT:	Physical Training
QRF:	Quick Reaction Force

REFORGER: Return of Forces to Germany
RIF: Reduction in Force
ROAD: Retired on Active Duty
ROTC: Reserve Officers' Training Corps
RSM: Regimental Sergeant Major (British)
SECARMY Secretary of the Army
SECDEF Secretary of Defense
SIDPERS: Standard Installation and Division Personnel Reporting Systems
SOUTHCOM: Southern Command
TSC: Theater Support Command
TTP Tactics, Techniques and Procedures
TRADOC: Training and Doctrine Command
UN: United Nations
UNC: United Nations Command
USA United States Army
USAF United States Air Force
USFK: United States Forces—Korea
VC: Viet Cong
WAC: Women's Army Corps

Staff Designations: G for Division and Higher; S for Brigade and Battalion

G–1/S–1   Personnel and Administration
G–2G–2/S–2 Intelligence and Security
G–3/S–3   Operations and Training
G–4/S–4   Logistics and Supply
G–5/S–5   Plans
G–6/S–6   Signal and Communications
G–6/S–7   Training

About the Authors

SMA (Ret.) Julius W. Gates

Julius W. Gates, a native of North Carolina, served 33 years in the U.S. Army, including tours in Germany, Vietnam, and Korea. As the 8th Sergeant Major of the Army, he championed training initiatives like the Year of Training and Year of the NCO campaigns, which led to the creation of The NCO Journal. Gates was instrumental in developing the Battle Staff NCO Course, a new NCO Evaluation Report, and the Self-Development Test. He continues to support the Army by advising the Joint Readiness Training Center and visiting Soldiers in the field.

COL (Ret.) Wes Martin

Colonel Wes Martin is a decorated U.S. Army veteran with over three decades of leadership experience in military operations, law enforcement, and international security. A pioneer in counter-terrorism and force protection, Colonel Martin has served in key positions across the globe, including Iraq, where he oversaw coalition base security and managed high-stakes negotiations. His expertise spans strategic planning, risk assessment, and crisis management, making him a sought-after advisor and speaker on global security and leadership. Today, Colonel Martin continues to inspire others through his dedication to service, resilience, and the pursuit of excellence.

Index

A

Abrams, Creighton Jr. 1, 157, 158, 159, 161, 209
Abrams (tank) 191, 209, 210, 241, 262, 263
Adair, Paul N. "Red" 284
Andrews, Agnes 6, 7
Armstrong, James E. 178, 182, 187, 188, 189, 193, 194, 195, 196, 246

B

Barajas, Tino 46
Barnett, Ross 47, 48, 49
Battle of Ia Drang. *See* Ia Drang
Be All You Can Be viii, 1, 148, 265, 304
Body count 76
Bolshevik Revolution 124
Bradley Fighting Vehicles 263, 283, 293
Bradley, Omar 1, 220
Bush, George H. W. 260, 265, 266, 267, 269, 274, 278, 282, 284, 289, 290

C

Callaway, Howard "Bo" 155
Carpenter, William "Bill" 38, 41, 46, 71, 72, 74, 75, 77, 78, 79
Cater 39, 40, 41, 42, 45, 46, 49, 50, 51
Cater, Richard J. 39, 40, 41, 42, 45, 46, 49, 50, 51
Checker-boarding. *See* Recondo Tactics
Cheney, Richard 266, 268, 269, 282
Crowell, Howard 197, 199, 200, 201, 202, 203, 205, 206, 299, 300

D

Douglas, Ethel Lee 2
Dunaway, George 298

E

Eisenhower, Dwight D. 5, 8, 18, 47, 48, 63

Emerson, Henry E. "Hank" 45, 292, 293
Endara, Guillermo D. 265, 267

F

Faubus, Orval E. 47
Ford, Henry 4
Franco, Victor 42, 43, 46, 95, 155, 205, 291, 301
Franks, Frederick M. Jr. 279, 280
Fulda Gap 176

G

Gates, Alice 3, 6
Gates, Lauren 136, 245, 246
Gates, Margaret xii, 59, 60, 61, 94, 96, 106, 107, 121, 123, 134, 135, 136, 137, 174, 212, 213, 214, 215, 224, 225, 226, 229, 239, 240, 241, 243, 245, 246, 298, 299
Gates, Melissa 60, 96, 121, 123, 136, 174, 243
Gates, Thomas 3, 4
Geneva Convention 88
Giap, Vo Nguyen 23, 116
Gorbachev, Mikhail S. 259, 260, 263

H

Helmstedt Support Detachment 189
Highway of Death 284
Hitler, Adolf 5, 21, 123
Ho Chi Minh 116
Hussein, Saddam 272, 273, 274, 275, 278, 279, 280, 281, 282, 284, 287, 289, 290

I

Ia Drang 62, 77, 78, 114, 306

J

James, Jesse 71, 89
Johnson, Lyndon B. 62, 63, 87, 88, 107, 191

K

Komitet Gosudarstvennoy Bezopasnosti (KGB) 260

L

Livsey, William J. 229, 230, 232, 234, 238, 240, 241, 242, 244

M

Marshall, S.L.A. 55, 56
Marsh, John O. 254
McClellan, George 167
McDade, Robert 77
McNamara, Robert 62, 63, 87
Meredith, James H. 47
Montagnards 68, 83
Moore, Hal 62, 77, 78, 79, 114, 306
Morrell, Glen 158, 162, 163, 215, 217, 218, 221, 226, 227, 228, 230, 245, 246, 299
Morrell, Karen 215, 226, 299

N

Noriega, Manuel A. 264, 265, 266, 267, 268, 269, 287
North Atlantic Treaty Organization (NATO) 191, 192, 198, 260, 262, 303, 308
Northern German Plain 176, 187
North Vietnamese Army (NVA) 23, 67, 68, 69, 70, 71, 72, 73, 74, 75, 76, 79, 81, 85, 86, 88, 108, 109, 111, 112, 114, 115, 116, 308

O

Operation
  Desert Farewell 286
  Desert Shield xi, 275, 297
  Desert Storm 209, 278, 285, 287, 289, 297
  Just Cause xi, 266, 269, 270, 271, 297

P

Panama Canal 265, 266
Patton, George IV 128, 129, 131
Patton, George S. Jr. 79, 117, 128, 129, 131, 159, 215, 255, 282
Paz, Robert 265
Pershing, John J. 20, 215
Plato 271
Plumley, Basil L. 78
Powell, Colin L. x, xi, 1, 252, 253, 265, 266, 275, 276, 278, 279, 289

R

Reagan, Ronald 188, 260, 263, 264, 266, 270
Recondo School 45
Recondo Tactics 45, 292
Rogers, Robert 304
Rogers, Will 263
Roosevelt, Franklin D. 150
Roosevelt, Theodore Jr. 265

S

Schwarzkopf, Herbert N. Jr. "Norman" 273, 275, 276, 279, 281, 282, 283
Skinner, Bill 46, 60
Stalin, Joseph V. 17, 18, 124
Strategic Defense Initiative 263
Sullivan, Gordon 254, 261, 266, 272

T

Tet Offensive 23, 116
Thurman, Maxwell R. "Mad Max" 265, 268, 269
Training the Force (FM 25-100) 248, 254
Truman, Harry S. 17, 18, 288
Truscott, Lucien 198, 211
Tully, Robert 77

V

Valley Forge 65
Viet Cong (VC) 68, 70, 82, 83, 86, 99, 114, 115, 309
Villa, Pancho 215
von Steuben, Friedrich Wilhelm 261
Vuono, Carl 244, 245, 246, 247, 248,

249, 251, 252, 253, 254, 257, 259, 260, 261, 262, 266, 268, 269, 271, 272, 276, 277, 278, 282, 283, 284, 285, 287, 288, 296, 297

W

Warsaw Pact 49, 124, 176, 189, 192, 198, 199, 211, 242, 260, 263
Weinberger, Caspar xi, 266, 275, 278, 289
Weinberger-Powell Doctrine xi, 278, 289
Weyand, Fred C. 23, 24, 127
Williams, Samuel T. 58
Wooldridge, William "Bill" 258, 298, 299

www.ingramcontent.com/pod-product-compliance
Lightning Source LLC
Chambersburg PA
CBHW050327010526
44119CB00050B/702